Copyright © 2008 by Robert B. Laughlin
Published by Basic Books,
A Member of the Perseus Books Group

Books published by Basic Books are available at special discounts for bulk purchases
in the United States by corporations, institutions, and other organizations. For more
information, please contact the Special Markets Department at the Perseus Books Group,
2300 Chestnut Street, Suite 200, Philadelphia, PA 19103, or call (800) 810-4145, ext.
5000, or e-mail special.markets@perseusbooks.com.

Library of Congress Cataloging-in-Publication Data
    Laughlin, Robert B.
    The crime of reason : and the closing of the scientific mind / Robert B. Laughlin.
       p. cm.
    Includes bibliographical references and index.
    ISBN 978-0-465-00507-9 (hardcover : alk. paper)
    1. Freedom of information. 2. Information policy. 3. Academic freedom.
4. Communication of technical information. 5. Intellectual freedom. 6. Knowledge
management. I. Title.

JC598.L38 2008
323.44'5—dc22
                                                                    2008024592

10 9 8 7 6 5 4 3 2 1

CONTENTS

1

THE END OF INNOCENCE

When we are young, we learn that knowledge is a beautiful, logical thing that anyone can use as he likes—provided, of course, he has the patience to read and think. This idea partly comes from parents, who never tire of inventing reasons for us to study more, excel in exams, and so forth, but it's also something we usually conclude on our own. Most of us decide in young adulthood that the ability to reason and understand is natural, human, and rightfully ours.

Unfortunately, this conclusion is erroneous. While some information is indeed available for free and even forced upon us in school, most economically valuable knowledge is private property and secret.

The owners of this knowledge do not want it made public and certainly do not want the state paying people to "discover" it. One can argue endlessly about whether "no trespassing" signs in libraries and schools are good things, but the debate is academic. As a practical matter, our rights to learn have already been circumscribed.

People often have trouble speaking about this problem because it's a worldly matter—like the practicalities of having children—that polite individuals don't discuss. Instead they smile and insist that education is golden and that the various ways of withholding knowledge—intentional generation of confusion, stonewalling, lying, disinforming—are obnoxious but not conspiratorial. They then deflect the discussion in a new direction by declaring the concerned person to be paranoid.

This denial is extremely irresponsible. The issue is the criminalization of learning. It's important. It's something we need to think about.

Our tendency to underestimate the danger of knowledge sequestration is partly a side effect of our otherwise sensible practice of separating knowledge into "technical" and "nontechnical" categories, rather like social class divisions, and then dismissing sequestration of the former because it's unimportant. Unfortunately, the reasoning behind this practice is exactly backward. We don't really accept sequestration of knowledge because it's technical. We redefine knowledge to be technical when it becomes sequestered. That is to say, when intellectual activity becomes valuable enough to be bought and sold, it changes character. The owner doesn't bother to explain it clearly to you anymore, and you don't bother to ask him for details. You just buy his product—or not, as you choose. That's why repairing your car is technical while driving your car is not. Both involve knowing how cars work, but repair is something you can buy relatively cheaply on the open market, whereas getting where you want to go is expensive unless you do it yourself. On purely intellectual grounds, however, there is no difference between technical and nontechnical knowledge.

Once you accept that the issue might be economic rather than cultural, you are forced to rethink some legal fundamentals. Refusing to apply traditions of free speech and free inquiry to things because they're property is different from doing so because they don't matter. A tiresome discussion about technical detail suddenly becomes a deadly serious one about the conflict between personal freedom and property rights.[1] The freedom in question is not the familiar one of political speech, instituted to discourage the consolidation of power by governments, but the freedom to learn and understand things relevant to making your living. In the past, nobody worried about protecting this freedom because the major problems of the day were political, and technical property issues didn't obstruct personal economic betterment.

But now we have entered the Information Age, a time when access to understanding has become more important, in many instances, than access to physical means.[2] The growing efforts of governments, corporations, and individuals to prevent competitors from knowing certain things that they themselves know has led to a stunning expansion of intellectual property rights and the strengthening of state classification powers. The Digital Millennium Copyright Act of 1998 and EU Copyright Directive of 2001, for example, make it a crime to circumvent anti-piracy measures (understand encrypted communication) and distribute code-cracking devices (tell other people about it).[3] The Bayh-Dole and Stevenson-Wydler Acts of 1980 redefine the mission of government-supported research to be the generation of intellectual property.[4] The Microsoft antitrust decision institutionalizes corporate monopolization of communication.[5] Courts now sustain patent claims for hiring strategies, real estate sales techniques, the discovery of chemical correlations in the body, and gene sequences.[6] Broad areas of two sciences, physics and biology, are now off-limits to public discourse because they are national security risks.[7] Our society is sequestering knowledge more extensively, rapidly, and thoroughly than any before it in history. Indeed, the Information Age should probably be called the Age of Amnesia because it has meant, in practice, a steep decline in public accessibility of important information.[8] This is particularly ironic given the rise of the Internet, which appears to spectacularly increase access to information but actually doesn't.[9]

The attitudes about knowledge implicit in this development raise profoundly troubling questions about human beings' fundamental rights to question and know. More and more, the "flash of insight" that we so admired in Galileo and Newton—the sudden understanding of a thing and its implications—is turning out to be a patent infringement or a state security danger.[10] More and more, the act of reasoning something out for yourself is potentially a crime.

The increasingly conservative legal interpretation of invention as theft echoes our society's growing ambiguity over how it feels about technical power. We sympathize with the young genius who, in an impetuous act of reason, breaks through the confusion and makes a glorious contribution to knowledge.[11] We also fear the genetic manipulation, nuclear conflict, usurpation of airplanes by terrorists, job export, and so forth that his contributions might facilitate. Unable to decide which is more important to us, we label his acts as criminal, or not, after the fact, according to principles that shift over time and that he didn't understand when he did his work.[12] He is like the soldier making brave decisions on the battlefield without knowing whether he will receive a commendation or a court-martial. We respect what he stands for but absolutely will not grant him full creative license. Too much is at stake. The irresponsible publication of a trade secret or military technology "discovered" by accident could mean death for a corporation, chaos in the streets, or loss of life in war.[13]

Thus at the dawn of the Information Age we find ourselves dealing with the bizarre concept of the "crime of reason," the unsocial nature or outright illegality of understanding certain things. Legislatures, with our tacit blessing, have begun writing laws that criminalize understanding and speech because it is easier than criminalizing the behavior they engender.[14] The argument they make, echoing that of previous eras, is that incremental curtailment of freedom is a reasonable price to pay for continued safety and prosperity. We will regulate and censor certain things for your own good. Don't worry about the details. They're technical. But who will the censors be? To whom will they report?[15]

Unfortunately, the simplistic reflexive response technically informed people make, "liberty or death," falls on deaf ears. It simply isn't workable. Our society has already decided, quite firmly, that a growing body of technical understanding shouldn't be accessible to everybody.[16] We have no option other than to sit down and plan, as best we can, what the rules of knowledge containment will be. That requires informing ourselves of the facts and thinking hard about them, since things you understand incompletely are easy to dismiss as confusing, boring, and irrelevant, even when they aren't.[17] Sanitized knowledge is also deliberately designed to look this way.[18]

Meanwhile the wiser heads are sorrowful and silent, for they understand the full significance of this moment. It marks the final, terrible demise of Enlightenment optimism. Descartes' brave declaration, "I think, therefore I am," has become a satire. We have collectively resolved to relinquish our intellectual rights, to vote them out of existence on the grounds that they are too inconvenient and frightening to live with. The "technical" nature of the banned knowledge is irrelevant. Knowledge is knowledge. Once we accept that some of it is too important for ordinary people to have, we are no longer at Orwell's doorstep but sitting together in his parlor discussing proper placement of the furniture. That's not the way many of us wanted it, but that's the way it is.

2

DANGEROUS KNOWLEDGE

Knowledge is dangerous. We wish it weren't so, and we like to think we're safer because we know more than our ancestors did, but we're just fooling ourselves. Our homes, workplaces, and social environments are bristling with potentially harmful technologies that we control only with great effort. The burden of remembering how to avoid accidents gets worse every day. Assaults on our peace of mind from people who know important things we don't materialize faster than we can deal with them. Far from being an anomaly of modern life, dangerous knowledge is everywhere.

There are more threats around than most of us notice. For example, you don't think twice about the butcher knives in your kitchen, yet knowing how to use them can, and routinely does, lead to tragic accidents and murder.[1] You don't think much about matches, yet knowledge of how to light them leads routinely to burn trauma and arson.[2] Knowledge of how to control rats leads to poisoning. Knowledge of baseball bats leads to smashed skulls. Nail guns, chain saws, tar vats, pruning shears, bathtubs, lace curtains, knitting needles—there's an endless list of potentially lethal instrumentalities in our homes that we casually accept because we've learned to accommodate their dangers.[3]

The abundance of dangerous knowledge in our lives is not a coincidence or a devil's plot but simply a side effect of economic activity. As human beings learn to make their way in the world, they automatically acquire the power to harm things, including each other. It isn't possible to legislate this power away, because making value for someone else requires manipulating your surroundings in ways they cannot. If you want, you can fill your head with knowledge that couldn't possibly be dangerous, such as telephone numbers or shapes of grains of sand, but you'll soon be unemployed if you do. Everyone knows this. If you want to survive, you had better acquire knowledge that empowers you and is therefore potentially dangerous. That's what other people want to buy.

Fortunately, to be safe, we don't have to ban all such knowledge. We can instead exploit the principle of safety in numbers. This strategy works fairly well because humans are, for the most part, smart enough and responsible enough not to hurt each other with the dangerous knowledge in their heads. For example, people misuse kitchen knives so infrequently that we don't bother to regulate them, even though the occasional misuses that do occur are ghastly.[4] Fireworks, by contrast, cause harm fairly easily, so we do regulate them.[5] Banning the sale of fireworks doesn't stop their use, however, since some people defy the ban, but it does discourage use and so reduces our risk of injury. In either case, we become "safe" because the likelihood of getting hurt is acceptably low, not because it is zero.

Where you draw the line between acceptable and unacceptable risk is fundamentally an economic calculation, not a political one. This is something people often misunderstand when they lobby their governments to ban this or that dangerous thing. For example, if someone tried to ban kitchen knives because they were too dangerous, everyone would laugh and use them anyway. They are just too useful. If someone tried to ban ammonium nitrate and diesel fuel, the chief ingredients of terrorist car bombs, black markets would quickly develop for both.[6] Farmers facing bankruptcy would find a way to get the fertilizer they need. Truckers needing fuel would do the same thing, although perhaps not as gently. The only significant effect of the bans would be increased prices for fertilizer and fuel. The real reason dangerous military technologies such as laser-guided bombs, cruise missiles, tanks, and antiaircraft guns aren't in stores is that they aren't useful for making money. Banning has nothing to do with it. Genuinely useful military technologies, such as the Internet or the Global Positioning System, have historically proved impossible to contain.[7]

A familiar example of enthusiastically embraced dangerous knowledge is driving. People learn to drive for lots of reasons, but it's typically because they can't get to work, fetch food, deliver construction materials to their job site, and so forth unless they do. We joke about how nerve-racking driving is, but it isn't a joke. One timing mistake can get you rammed. One speed misjudgment can crash you into a concrete porch. One overlooked visual clue can kill a pedestrian. We've all seen vehicles used as a weapons in low-budget television shows, Jackie Chan movies and the like, but the threat isn't just Hollywood make-believe. One suicidal driver could take out fifty people—more if he's driving a truck—by intentionally crashing through a busy highway divide into oncoming traffic. Multiply that by the number of potential suicide murderers who can drive and you get a threat to life much greater than that from airplane explosions. Even without murderous intent, driving kills a lot of people. In 2005, traffic accidents dispatched 42,000 in the United States, a comparable number in Western Europe, and 1.1 million worldwide.[8] That's three hundred times the number of people we lost in the 9/11 attacks. Knowledge of driving is dangerous.

Not only do we live comfortably with lots of dangerous knowledge, we're genetically programmed to seek it out. Young mothers quickly learn how adept their toddlers are at turning limited experiences into experiments that could kill them, from walking out in traffic to poking sticks into electric sockets to eating interesting things that aren't food. When the kids get a bit older, they advance to climbing trees, which they find irresistible, despite the obvious and potentially lethal danger of falling out. Boys cannot be near rocks without throwing them, no matter how sternly and repeatedly you lecture them about brain hemorrhages. Taking the rocks away just diverts their attention to sticks, which they turn into guns, often with nuclear bullets, and use to shoot each other. The obsession with dangerous things keeps worsening until you get to bungee jumping, hang gliding, and skiing *way* too fast, although by that time it's not your problem anymore.

Our strategies for controlling even more dangerous "industrial strength" knowledge are also largely economic. For example, you don't worry about an irate neighbor seeking revenge by crushing your car with a huge earth-moving machine. The reason is that earth-moving machines are expensive and difficult to get. You also don't worry about his paving over your front yard with blacktop, filling up your pipes with sulfuric acid, or blowing up your house with dynamite. He could read up about these things in the library and learn how to make all the parts himself from cheap, accessible ingredients—tar, sulfur, nitrogen, soap, and so forth—but actually doing so would take him so long that it's effectively impossible. You occasionally encounter people with the financial resources to hire an asphalt company, telephone for a few tank cars of acid, or order up a few boxes of dynamite from their mines, but such encounters are rare. Moreover, as in encounters with people wielding knives, you're just not going to be safe in those rare moments. If your country's top general appears at your door with some of his boys, you let him in. If your boss suggests that you not reveal any of his multinational company's secret formulas, that's what you do. If the local Mafia don complains that your dog Rover yaps too much, you shoot Rover.

The principle of economic control even applies to nuclear technology. We tend to think of nuclear matters as occupying a special danger category all their own, but they don't. Nuclear reactions release about a million times more energy than chemical ones do,[9] and that does indeed make them a million times more dangerous.[10] However, a million times fewer of them are required to keep your house warm. Moreover, a nuclear engine or battery could, in principle, be a million times smaller than a conventional one delivering the same power.[11] Were it possible to make small nuclear power plants, no amount of government restriction would stop people from making piles of money powering cars, airplanes, and portable electronic gizmos with them. But, alas, it isn't possible. The machines that convert nuclear reactions into electricity cannot be miniaturized, because great mass is required to capture and tame reaction products flying off at high speeds. This technical problem, and nuclear power's consequent impracticality in the consumer marketplace, is the real reason nuclear technology hasn't proliferated.

Nonetheless, nuclear knowledge is extremely dangerous. The reason is precisely the one you see in nuclear horror movies: nuclear weapons, unlike nuclear power plants, are small enough to fit in a backpack.[12] That effectively gives a single individual a million times more destructive power than one with a conventional bomb. The factor of one million makes the explosion useless in ordinary commerce but, sadly, highly valuable in war.[13] Moreover, this lopsided value proposition is permanent. You can't make tiny nuclear weapons for the same reason you can't make tiny nuclear power plants. It's all or nothing.

The special combination of commercial uselessness and extreme military effectiveness has led us to invent a new, quasi-military way of containing the danger of nuclear technology: we ban the knowledge itself.[14] This strategy is modeled after the practice of classifying weapon designs as state secrets and regulating communication about them with espionage laws.[15] Just as you can go to jail for disclosing design information about aircraft carriers, torpedoes, or fighter planes, so can you go to jail for disclosing nuclear secrets. But there is an important difference. It isn't just the blueprints of nuclear bombs that are classified but the *physical principles* on which they are based. The Atomic Energy Act of 1954 imposes sweeping restrictions on what you can say about nuclear energy in public, including disclosing what knowledge is classified.[16] Under this law, you can go to jail for discussing the results of certain calculations that anybody could do in his spare time, even though these calculations are not explicitly forbidden by any publicly accessible document. Moreover, while conventional military classification is temporary because hardware designs eventually go obsolete, nuclear classification is forever. In the interest of national security, we have made an entire branch of human knowledge disappear.

Of course, the danger itself didn't disappear. That's why you can't use nuclear classification practices as a handy recipe for containing dangerous knowledge. Classification merely sets the price of obtaining know-how so high that amateurs and budding entrepreneurs can't get started. States, on the other hand, have virtually unlimited amounts of money, so many of them have all the allegedly secret nuclear knowledge at their fingertips—the recent revelations about nuclear activity in Pakistan and North Korea being cases in point.[17] Classification didn't impede them at all. States also have the resources to make fissile fuel, buy special materials, machine delicate parts, and so forth, the cost of which is the barrier to proliferation that actually counts.

Since the effects of classifying nuclear knowledge are, at best, indirect reinforcements to the vastly more significant barrier of cost, nobody knows whether classification will matter in the long run or not. So far, states that have successfully turned nuclear knowledge into actual weapons (including those possessing nuclear weapons but denying it for political reasons) have been responsible about using them because of their own vulnerability to retaliation. But their responsible behavior could end very quickly if one country decided to attack another in secret by slipping nukes to terrorists.[18]

Unfortunately, the illusion of nuclear classification success has beguiled and misled a lot of people. That, in turn, has caused the banning of certain kinds of thinking and speech to insinuate its way into laws regulating other kinds of dangerous knowledge— with predictable results.

Take cryptography. The need to communicate without being understood by the enemy has always been essential in war, so it's not surprising that encryption— the garbling of a message so that it can't be understood except by the desired recipient—is a highly guarded military secret in most countries.[19] But unlike nuclear energy, knowledge of encryption is commercially useful, in particular for keeping financial transactions secret.[20] Thus it was only a matter of time before this know-how escaped into the private economy.[21] Exactly when this happened is difficult to pinpoint, but it began to genuinely alarm governments when the Internet took off.[22] They suddenly realized that agents of hostile states, industrial spies, criminals, and terrorists could communicate securely at blinding speed while hiding their communication in plain sight among the babble of other encrypted communication on the Net. Not only couldn't you spy on the criminals, you couldn't even tell that they had communicated! That led to a flurry of laws in various countries intended to curtail the spread of encryption "technologies." All of them failed.[23] One reason they did is that encryption is not a "technology" but a handful of simple mathematical principles that anyone can understand. In this sense, the laws effectively tried to make two plus two equal five. The more important reason is that the legislators were trying to defy the laws of economics. Banks must have ways of communicating with their customers that governments can't oversee. If they don't, the customers will take their business elsewhere.

The idea of criminalizing encryption knowledge because it's dangerous hasn't disappeared. It has just changed its name and allied itself with copyright law.[24] We have all read about the armies of young people around the world who use their computers to copy and share music and movies that, in previous times, they would have bought.[25] Prosecuting each of these copyright violators would be prohibitively expensive, so content providers have obtained laws criminalizing the knowledge of how to copy. The laws don't say that, of course, but it's what they amount to. They make it a crime to disseminate "technology" that facilitates the evasion of copy protection.[26] However, the "technology" is just decryption, which is the same thing as understanding. The threat to society from file sharing is somewhat clearer in this case, since this practice could bring down billion-dollar industries if not curtailed. But thinking about the problem in terms of nuclear threats misses the point entirely. Using machines to speed up understanding a million-fold may indeed be evil, but that doesn't matter. What matters is that the knowledge of how to share files economically benefits more people than it harms. That makes it very difficult to stop.[27]

The proper analogy for copyright infringement is probably not with nuclear weapons but with recreational drug use. We criminalize the latter because of its extremely detrimental side effects, not because it poses a military danger. We calculate that the social value of reducing drug use outweighs the inconvenience of restricting certain economic activity. However, drug laws are notoriously ineffective at actually stopping drug use, especially in affluent communities.[28] People just defy the law and then hush up problems as necessary. Singapore, which executes drug possessors, prides itself in being a counterexample,[29] but public discourse in Singapore is so controlled that it's difficult to know how warranted the pride is.[30] Moreover, the argument cuts both ways. Becoming more like Singapore is precisely what criminalizing knowledge and public discourse in the name of safety is all about.

The knowledge issues that scare us the most are biological.[31] Hardly a day goes by that you don't read about Marburg virus, genetically engineered smallpox, chimeras, Frankenfood, cloning, or some other unsettling biotech surprise.[32] Scarcely a day goes by that you don't hear the corresponding cry of alarm. Scientists will endanger the food supply.[33] They will enhance disease.[34] They will kill fetuses.[35] They will make monsters.[36] They will obviate motherhood.[37] They will destroy Earth.[38] Our fears are partly subconscious and thus sometimes a bit irrational. Everyone remembers the plot of *Frankenstein*. Everyone remembers the end of H. G. Wells's *War of the Worlds*, where bacteria beat the Martians after tanks and guns could not.

Yet biological threats are very different from potentially classifiable engineering ones. Biological "technologies" are part of the economy already and so can't be prevented from escaping there. The important disease organisms, for example, have coexisted with humans for millions of years. Scientists can modify these organisms to make even deadlier ones, but it's difficult to imagine what would be gained by making something more horrible than ebola, cholera, or plague.[39] Also, humans have been modifying their own genes efficiently for millions of years by surviving into adulthood (or not) and choosing whom to marry. There is nothing new about genetic modification other than the involvement of chemists, which parents who know something about chemists find very amusing. There also isn't much point in having babies in test tubes as long as having them the old-fashioned way remains cheap, easy, and entertaining.

Biological knowledge is nonetheless extremely dangerous. Immediately after the World Trade Center attacks, letters containing anthrax spores were mailed to news organizations and to the U.S. Congress, killing five people and injuring another seventeen.[40] The perpetrators were never identified. Only eighteenth-century technology was involved—the mail and a common disease bacterium. The lesson was not lost on anyone that the number of victims could just as easily have been 100 million. Equally frightening scenarios have been imagined for bioterror attacks using smallpox virus and botulinum toxin.[41]

In response to these fears, there are now concerted efforts to make dangerous biological knowledge more difficult to get. While no "born classified" principle is at work, we are increasingly seeing "voluntary" decisions not to engage in certain research activities or publish certain findings, especially in matters potentially relevant to war.[42] Biological knowledge is not becoming classified in the strict sense of the term, but its dangerous parts are slowly becoming taboo, just as the dangerous parts of nuclear technology did a half century before.

The special immediacy of biological fears (everyone has been sick) has even made many people rethink the proposition of generating knowledge for its own sake. There have been new admissions that knowledge can hurt us,[43] thoughtful suggestions that some knowledge should not be created in the first place,[44] and furious complaints about release of certain knowledge to the public.[45]

This hand-wringing is somewhat excessive, however, in that biological technology, while more tangible, isn't necessarily more frightening than other kinds of technology. The 1995 Tokyo subway attacks, for example, which involved only chemistry, killed and injured more people than the anthrax mailings did in 2001.[46] A nuclear attack hasn't happened yet, but that notoriously troubling threat has made its way into popular culture.[47] The Comprehensive Test Ban Treaty is the legal prototype for the recent United Nations ban on human cloning.[48]

Moreover, the sad truth is that preventing dangerous knowledge from being created is impractical. The knowledge itself is too useful, and anyway it's generated mainly in private for specific industrial or military purposes and is thus not completely under popular control. Public domain scientists like to brag about how potentially dangerous they are, but it's mostly bluff. Scientists without money aren't actually very dangerous. Thus while pundits are indulging in fantasies about stopping knowledge creation, more responsible parties are thinking hard about history, economics, self-interest, and coping strategies. It's like terrestrial comet impacts.[49] It doesn't hurt to talk about banning comets, but it's a good idea to think realistically about comets, correctly assess their threat, and plan what to do if they hit.

3

THE MASTER CRYPTOGRAPHER

Albert Einstein, in a pique over his colleagues' wanton theoretical liberties, said that God does not play dice with the universe.[1] What he meant was that quantum mechanics, a theory that they were developing, couldn't be right because it required randomness to result from attempts to measure things. It made no sense to him that simply looking at a tiny domino would make it topple, one way or the other, in a direction you couldn't determine ahead of time. The controversy later turned out to be a miscommunication over what "measure" meant, and so it became a kind of joke that was relegated to the footnotes of quantum mechanics books. But Einstein's remark also became an authoritative statement of the belief, held by physicists to this day, that the universe unfolds logically.

It isn't true, of course. God does play dice with the universe, and He has done so routinely from the beginning.[2] He created the galaxies in a huge disorderly explosion. He peppered the stars randomly in the sky. He appointed the waters of heaven to descend as rain or snow in undisciplined hordes. He ordered dry land to meet the seas at beaches composed of sand grains of dazzling variety. He persuaded the waves to wash upon them in lovely patterns of shape and sound that were never the same twice. He arranged for the summer breezes to blow haphazardly and, in moments of calm, for flies to zip around unpredictably so that they couldn't be swatted. And in an inspired burst of genius, He invented people, who made dice for Him and played with them, and then went to work writing scholarly documents declaring that God doesn't play dice with the universe.

Einstein must have been preoccupied, for he overlooked the obvious difficulty that the tumbling of dice isn't actually random but *chaotic*: perfectly deterministic and logical yet so unruly that it's effectively random.[3] The unruliness of dice comes from their occasional tipping onto corners, where tiny differences in attitude can change the direction of fall and thus have a profound effect on the entire subsequent history. If you try to predict the outcome of a throw using a computer, you discover that errors you unavoidably make at these moments, no matter how small, grow to catastrophic sizes after a few tumbles. This happens because they expand by percentages, the way compounding bank interest does, rather than just addition. If you roll long enough, you will overwhelm the ability of any computer, or groups of computers, that could ever be built to correctly predict the outcome. In other words, chaotic things are unpredictable in the literal sense that human beings can't predict them. The fate of the dice is known after they leave the gambler's hand, but it isn't known to us.

Chaos isn't just a physical phenomenon. You also find it in pure mathematics.[4] Take out your hand calculator. Enter the number ½. Take its square root. Multiply by 10,000. Zero out all digits left of the decimal point. Take the square root again. Multiply by 10,000 again. Zero out all digits left of the decimal point again. Keep doing this for one hundred cycles and you will generate a series of one hundred decimal numbers between 0 and 1 that is, for all practical purposes, random. It isn't actually random, of course. You generated it logically and systematically using your calculator, and so you know exactly what the next number in the sequence will be at any moment. An uninformed onlooker doesn't, however, because he doesn't know the trick.

These simple examples and others like them—leaves fluttering in the wind, water splashing in a stream, Ping-Pong balls bouncing together down wooden stairs—are more than technical curiosities. They are beautiful prototypes for knowledge disappearance. They show how facts about the status of a thing, good as gold when first acquired, can degrade over time. They show how perfectly logical theories can become 100 percent wrong if extrapolated too far into the future. They show how some kinds of knowledge can be more important than others—how failure to know key things, whatever they are, can prevent you from putting a puzzle together, no matter how much inferior data you amass. In other words, they show that knowledge disappearance is not imaginary or an unfortunate consequence of your own stupidity but a commonplace natural phenomenon.

The nontechnical version of knowledge degradation is sadly familiar to most of us. It's the basis of the famous party game in which you whisper a rumor into someone's ear, let it travel from person to person, and see how it changes.[5] After about thirty such transfers, the final version of the rumor usually has little resemblance to the original one. The knowledge degrades with each retelling because each person, in internalizing the information, isolates what he or she believes to be parts that don't matter and then retells the story with these parts changed, thereby making it more interesting to the listener. The listener then does the same thing, except with different criteria of importance. Over repeated retelling, the fabricated material slowly becomes elevated to status of truth—especially fabricated material that seems reasonable by commonly accepted norms.

The lack of fidelity in the party game isn't just a flaw of verbal communication. You find it in writing too, and for the same reason. Writing has the significant advantage of being readily witnessed by third parties, but you exploit this advantage only if you return to original sources. Freshly minted writing, even that generated by diligent scholars, contains ancillary material inserted to make things more readable—and which the reader has difficulty distinguishing from original content. No matter how innocent these additions are, they inevitably grow with repetition into big mistakes. A particularly notorious example of this effect is the internal inconsistency of the Pentateuch, which many scholars argue was caused by post-facto editing.[6]

Nor is technical knowledge immune to degradation, even though it's ostensibly precise. For instance, it can simply be wrong. The well-documented "knowledge" that high voltage in power lines causes cancer, for example, cannot possibly be true because even higher voltages are everywhere in your home.[7] You find them in picture tubes of television sets, static cling of laundry in dryers, and even in your own body after you stroll across a carpet. That's why you get shocks from doorknobs in winter. This allegation is actually a distortion of perfectly correct reports of cancer clusters near electric power facilities, the cause of which isn't known. There are lots of possibilities other than volts. Another example would the remarkable absence of mechanical problems in commercial jets.[8] This also cannot possibly be correct, since no other such complex machinery is ever that trouble-free.[9] The comfortably low coliform bacteria counts that small water companies report must certainly be, in some cases, disinformation cooked up to avoid inconvenient and embarrassing alarm.

Scientific conclusions are wrong so frequently that scientists can't resist joking about the problem. There is, for example, the famous article in the *Annals of Improbable Research* carefully documenting how tornadoes in the United States occur frequently in states with large numbers of mobile homes—and concluding that mobile homes cause tornadoes.[10] Jokes nearly as good range from pithy satires of actual research to outright silly things such as using an expensive spectrometer to compare apples and oranges.[11] Still, you can't beat torna-does for pure fun. Bernard Vonnegut's lovely "Chicken Plucking as a Measure of Tornado Speed" won him a coveted Ig Nobel Prize, the notorious Harvard spoof of the real Nobel Prize.[12]

Another way technical knowledge can lose its value is by becoming a needle in a haystack—an essential thing buried under mountains of nauseatingly accurate but ul-timately irrelevant detail. Numbers, like rumors, are not all equally important. If you're trying to make computer chips, for example, you need some familiarity with atomic structure, chemistry, quantum properties of crys-talline silicon, and so forth, but the thing you *really* need is a little cooking rule: always let a small amount of steam into your oven during baking.[13] Why you need this steam is a long story involving chemical bonding errors that inevitably occur when you oxidize silicon, but the point is that people couldn't, and didn't, guess it from knowledge of basic chemistry back in the 1950s. They discovered it serendipitously in the laboratory. Naturally, when you discover such a thing you don't share it with the world but keep it to yourself and race to corner the market. This is just what happened. If some bit of your trade secret leaks out, you counter by leaking disinforma-tion along the same lines in the huge flow of correct, but unimportant, technical communication that you publish routinely. That's one of the main reasons young industries publish so much. The river of trash information confuses your competitor and forces him to spend intellectual resources ciphering out trivialities, while the truly important thing slips through his fingers.

Hiding needles in haystacks is also the key stratagem of cryptographers.[14] Returning to the calculator example, let us imagine that instead of multiplying always by 10,000, we multiply either by 10,000 or 100,000, depending on our whim, keeping all other steps the same. This modification wouldn't be noticed by an uninformed onlooker, who would still see an apparently random sequence of decimal numbers between 0 and 1. But an informed person could work backward from any two successive numbers to figure out which choice we made. Thus we can send him an encrypted message! Eight such choices in succession define a "byte," the basic unit of signal transmission on the Internet. So, given adequate time and patience, we could send him, through a long series of apparently random numbers, a byte-encoded text message, photograph, pirated song, or any other digital communication, all in plain sight. Moreover, storing up all the transmitted numbers on a huge disc drive and searching the drive with an equally huge computer wouldn't help outsiders decipher the code. To do that they must learn or guess the *key* piece of information, the trick by which the message was encrypted.

Not only can you encrypt secret messages into apparently meaningless random numbers, you can encrypt them into legitimate communications such as photographs or voice transmissions.[15] What enables this trick, called steganography, to work is the human brain's ability to accommodate noise. For example, if it's snowing, you can still recognize a friend's face, provided he isn't too far away. The falling snowflakes occlude your view, but your brain knows they're just mistakes and so removes them and fills in the missing spots of the image with what should be there. In other words, it makes things up. That's why you don't necessarily throw away a photograph as hopelessly ruined if it's out of focus or smudged. It's also why you don't discard a voice recording of a spoken conversation just because it has traffic noise, refrigerator hum, or screaming children in the background. In real life, we deal with such corruption all the time, and we're good at it. So, to hide a secret message, you just make it look like background noise! You render a picture or voice transmission into a series of bytes and then modify a small fraction of them intentionally. A person who doesn't know what you did interprets the modifications as naturally occurring mistakes and ignores them. A person who *does* know what you did, on the other hand, retrieves the modifications, throws away everything else, and reconstitutes the secret message.

Your brain doesn't edit only images and sounds, of course. It also edits concepts, conversations, and news. That's why secret agents in movies are always sneaking messages into newspaper articles and inventing code phrases, such as "Blue fox hungers in the bat cave for Szechuan spicy sauce." It's also why explaining a genuinely new idea to someone is so difficult. You speak perfectly clearly, but your listener doesn't hear what you're saying. Instead he interprets unfamiliar things as mistakes and registers only corrected versions of them in his mind. Unless you deliver the idea carefully and repeatedly, it just bounces off, like a Nerf ball. If you make the even bigger mistake of attempting to convey several new ideas at once, he perceives you as a babbling lunatic. "That person is either inspired or crazy," he will say, telling himself that it's definitely the latter.

This tendency of our brains to interpret unfamiliarity as noise has the insidious side effect of causing small, abstruse bits of knowledge to disappear. What happens is that they get lost in the river of trash information that washes about us, and that requires great mental energy even to dismiss. In a sense this problem is nothing new, for most human communication has always been largely advertising and thus crafted to get face time (and money) from other people without confusing them with unfamiliar content. The new development is electronic delivery, in particular, television and the Internet, the cheapness of which has stupendously increased the amount of useless work our brains must do. As we struggle with this awful burden, the familiar things in our lives rapidly become more familiar, and unfamiliar things become less familiar. That's why everyone knows that $E = mc^2$ but very few people know exactly what it means or what experiments it describes. The latter knowledge has almost disappeared.

The disappearance of little details is especially relevant to knowledge sequestration because it's precisely in such details that important technical value lies. This is partly a circular statement, since the value of a thing depends on how much people will pay for it—that is, zero, if the knowledge is widely familiar. But there's a deeper reason. Nature, it turns out, is a master cryptographer. At the most fundamental level, the laws of physics are laid out in plain sight for everyone to see. Yet you can't generally predict things with these equations for the same reason that you can't predict the roll of dice: accumulating errors destroy your calculations.[16] Writing down equations for billions of atoms and trying to solve them by computer doesn't lead you to designs for cars or airplanes. What leads to designs of cars and airplanes is the discovery, by accident in an experiment, of collective principles of organization encrypted into these equations—for example, that metal atoms will, of their own accord, congregate, crystallize, and begin conducting electricity. That there should be such principles is not at all obvious, and thus something not generally possible to deduce from the equations themselves. The knowledge is therefore analogous to the special key you need for decrypting a message. If you don't know it, you can't anticipate what will happen, no matter how much inferior data you amass. But the very impossibility of deducing these principles from fundamentals makes them abstruse by definition. That, in turn, makes them especially susceptible to disappearance. There are exceptions in which they become widely used and familiar, but the people who discover such principles tend to discourage this outcome, for it makes their discovery economically worthless.

There are many excellent examples of serendipitous discoveries that became sequestered knowledge and then were partly leaked—so we know about them. The highly secret art of Damascus steel manufacture, for instance, was based on discoveries of carbon chemistry in iron made accidentally in forging experiments— plus the subtle stabilizing influences of impurities such as vanadium and molybdenum accidentally present in iron ore obtained from particular mines.[17] The manufacture of synthetic organic dyes, which began as an accidental modification of the coal tar derivative aniline in England, eventually became such important industrial secrets in Germany that the government prohibited the issuance of passports to chemists.[18] Xerography, which began equally humbly as an observation of light-induced electrical effects in sulfur, became a fierce battleground of industrial secrecy when competing manufacturers began substituting proprietary polymers for the all-important amorphous selenium photoconductors.[19] There are so many similar examples, from shipbuilding to glassblowing, ceramics, metallurgy, pharmacology, and other technologies stretching back through history, that you can't begin to enumerate them. In not even one of these examples was the technical development predicted. Each was a surprise implication of a discovery that someone made by accident.

The sequestered knowledge that concerns us most at the moment can't be so reliably documented, for the obvious reason that it's still invisible. Nonetheless, you can spot it easily through telltale signs of sanitization. Examine an important new technical development such as the blue light–emitting diodes on your car's dashboard, the flash memory in your digital camera, or cloning, and you find that key facts, the specific details of how you make things and how they work, are suspiciously difficult to find or are completely absent.[20] Your first thought is to dismiss the omissions as unimportant and to come back to the details later only if you need them. You have it exactly backward. The missing details are the important things, and the generalities and vague descriptions that you prefer actually have no economic value. Moreover, if you tried to figure out these details, you would find that you couldn't do so without expending lots of time and money rediscovering key undocumented facts. Similar strange omissions also occur in the scientific record of nuclear physics and, increasingly nowadays, genetics.

Thus modern knowledge disappearance is easy to underrate or deny because it is beautifully subtle and takes place right before our eyes. It is accomplished not through the kind of ham-handed censorship we know how to fight but through convincing us that important things are unimportant and tricking us into throwing them out. This strategy works more efficiently now

than it ever did before, because the flow of trash information has so vastly increased. In this sense, electronic
technologies such as the Internet are not instruments
of knowledge dissemination at all but agencies of
knowledge destruction. What makes such a bizarre
state of affairs possible is that God plays dice with the
universe.

4

GAMES OF CHANCE

Scientists are, as a rule, very poor gamblers. There are exceptions, of course, but the vast majority of them chose their profession because they love reason, and so they don't like squandering time and money playing games that logically require participants to lose. This brainy righteousness sometimes has amusing consequences. In March 1986, Las Vegas newspapers buzzed with rumors that the MGM Grand Hotel and Casino had suffered its worst weekly take in history—including the week of its terrible fire.[1] MGM had made the mistake of hosting a big physics conference. The scientists, it turns out, didn't care for neon cowboys, tiger shows, topless barmaids, and other distractions. In fact, they complained after returning home that these things had interfered with their concentration at seminars. Vegas cabbies got real mileage from this story—and presumably generous tips too. No Las Vegas hotel has ever invited the physicists back.

Not surprisingly, most scientists are also poor. They don't think of themselves as poor, but instead imagine that reasoning ability is the most important thing in the world and that they, being the most logical of all people, are therefore the most wealthy. But now and then they notice less-gifted individuals zooming by in their Lamborghinis on the way to vacation hideaways, restaurants, and parties, sometimes waving a cheery greeting as they pass. After this happens a few times, the scientists begin to realize something is amiss.[2]

The misconception they confront is that economic life is not about logic at all but about game playing and deception—the exact opposite of science.[3] Your objective isn't to make the world a safer place for reason and light but to relieve other people of their money as efficiently as possible without going to jail. Making money is not for Easter bunnies.[4] You don't share, you don't always tell the truth, and you don't tell people everything you know. If a customer shows interest in a car on your lot, you don't blurt out that the vehicle's delivery price was $16,000 or that new models are coming in soon. You keep this information to yourself and induce him to pay $23,000 through a bag of psychological tricks: We can't lower the price but we'll throw in floor mats and vacuum the car's interior at no extra cost to you. For this week only, we'll also give you a special deal on underbody anti-rust coating. We'll also sell you a special prepaid maintenance contract that will save you more money than reducing the price would. And just to avoid any bad feelings, we'll give you $1,000 over book value for your worthless old BMW. We'll definitely go straight to hell for this decision, but your peace of mind is worth it to us. Pay no attention to the interest rate on those loan documents! It doesn't matter right now. All that matters is how good you'll feel driving out of the lot in this baby.

The importance of deception in economic life is why poker is so much more popular than bridge or canasta. The latter involve psychology and tactics, but in the end they're about defeating the enemy with assets you have. Poker is about defeating the enemy with assets you don't have.[5] Your task is to make opponents think that you have more power than you actually do and then bluff them into capitulating unnecessarily. People find poker irresistible because it's a microcosm of their professional lives.[6] All of us eventually learn, sometimes through sad experience, that what counts much of the time is not diligence, hard work, or forthrightness but fiendishly masterful fibbing.

Poker-like deception isn't a pathology of economics but a practice central to its function. In morality tales or religious texts, things are valuable or not according to absolute principles that never change. In economics, by contrast, value is a fiction created by human beings for exchanging goods and services.[7] Its only meaningful measure is the amount of money that changes hands in a buyer-seller relationship. That's why it isn't just childish obnoxiousness when a seller pretends to have more demand for his product than he actually does and a buyer pretends to disbelieve him. Their dance of deception and eventual compromise literally defines the price of the thing and thus its value. People who refuse to play this game because it's immoral or illogical are demonstrating a woeful ignorance of human nature and, in extreme cases, consigning themselves to poverty. In economics, nice guys always finish last.

The bilateral ignorance implicit in this practice is so central to our lives that it reconstitutes itself if disrupted.

Suppose, for example, your customer or supplier learns the precise details of your business situation. He then can, and usually will, insist on terms unilaterally advantageous to himself. You might follow through with this unhappy transaction or you might not, but you certainly wouldn't do so again and again. You would instead find another supplier, sell something else, or switch your line of work entirely. Your family has to eat, after all. In a similar situation he would do the same. Faced with one-sided relationships, people eventually seek out new ones in which each party knows something essential that the other party doesn't. Over time, this process tends to restore even violently disrupted buyer-seller equilibria.

An important implication of this behavior is that universal access to knowledge is fundamentally incompatible with market economics. It isn't just that free knowledge is inconvenient for rapacious software and entertainment companies. It's that free knowledge is inimical to the practices of asset exchange and pricing essential for economic life. In fact, it's so inimical that enacting laws forcing access to knowledge just induces people to invent cleverer ways of making it inaccessible. Learning to read and write, for example, is now free in most countries but learning to read and write *well* is expensive. Computers that connect to the Internet are cheap but computers that connect to the Internet *securely* cost extra. Access to knowledge, it turns out, is like vitamin A: a little bit is essential but too much is poisonous.[8]

Many people, including those smart enough to know better, fail to understand that ordinary economic activity causes knowledge sequestration rather than the other way around. Thus, Internet guru Esther Dyson continues to preach that Web technology will soon make intellectual property obsolete—even as Bill Gates responds that restricting intellectual property rights is tantamount to communism, and James E. Rogan, former director of the U.S. Patent and Trademark Office, calls intellectual property rights "the underlying basis of a free market economy."[9] Meanwhile, ordinary citizens, oblivious to these controversies, continue circumventing and thwarting each other as though it were natural— which it is—and using the ostensibly revolutionary Internet to purchase penny stocks, monitor mortgage interest rates, and patronize proliferating Texas hold 'em Web sites.

The confused idea that secrets make the economy, rather than the reverse, becomes particularly serious when legislators use it to justify the criminalization of learning. The reasoning goes like this: some knowledge in a modern society must be secret; therefore knowledge is property; therefore acquiring this knowledge without paying is theft; therefore people who learn things without paying must receive fines and jail sentences. That's how we got, for example, the Economic Espionage Act of 1996, which makes it a federal crime to take trade secrets across state borders for economic purposes (Section 1832) and an even bigger federal crime to take these secrets abroad (Section 1831).[10] This is a sweepingly broad law because a trade secret is, for all practical purposes, anything the victim knows that no one else does, at least before the espionage takes place. While we don't yet know whether this law will have the desired effect, we do know that it will be massively violated. The reason is that the affected parties are engaged in high-stakes poker, a game in which seeing through the other person's deceptions and lying better than he does matters considerably more than the cards you hold. Not even one of these parties believes that stealing ideas is immoral or that criminalizing learning is good for society. The very idea is absurd.

The legal confusion surrounding intellectual property is especially problematic, unfortunately, because law in Western tradition isn't game playing or deception, at least in principle. We all know that there are frequent miscarriages of justice, even spectacular ones like the O. J. Simpson trial, but we're angered and ashamed by these failures of our legal system and want our governments to reduce them.[11] That's why gift shops frequented by lawyers and government building contractors do such a booming business in images of Lady Justice, the blindfolded faux Greek goddess holding scales in one hand and a sharp double-edged sword in the other. Cynics are sometimes inclined to dismiss Lady Justice as a kitschy joke, but they miss the point. Lady Justice is not a technical concept but a religious one. She embodies Logos, the idea from Greek Stoic philosophy that translates variously as Nature, Reason, the Universe, and God.[12] Logos comes down to us from Roman times through the Judeo-Christian ethic and is deeply embedded in most aspects of European civilization. The identification of law with Logos is why the legal system we demand, and largely have, is the exact opposite of economic opportunism. We don't want Lady Justice playing poker. We want her regulating our animal instincts with her sword and her power of Reason. The latter is so important to us that we just don't accept the authority of laws that are capricious and don't make sense. We simply disobey them without remorse. Such laws therefore cannot survive in the long run.

Thus the key element missing from much legal thinking about intellectual property is simply morality—a fact wonderfully consistent with the increasing popularity of tasteless lawyer jokes. All intellectual creations begin life as fruits of reason. That means they start out in a special category of purity all their own and only later become property by being used as bets in economic poker games. Like angels falling from heaven, they then change status. Labeling the creation of secrets as good because it protects inventors' rights is just ridiculous—like saying that card playing is moral because people have a natural right to lie. In reality both practices are fleshy corruptions that we put up with because they're necessary for living.

That these legal traditions try to serve God and mammon simultaneously makes it difficult to take seriously arguments for greater protection of commercial secrets on the grounds of righteousness. You're inclined to smile and imagine more worldly grounds: people usurping Lady Justice's sword to make more money for themselves at the expense of someone else.[13] The most celebrated example of this problem is the legal controversy surrounding Microsoft, but there are countless others.[14] The rhetorical sleight of hand occurs in the very last step, where you resolve to punish individuals who learn things without paying because their action is *wrong*. What's wrong with stealing ideas in a poker game? Stealing is part of the milieu, something you yourself engage in, and against which you routinely defend. If you fail to stop the thievery it's too bad—for you—but not something you're justified in complaining about. It makes no sense to label a thing wrong in an enterprise that is immoral in the first place.

The deep conceptual link between creations of the mind and the law itself gives intellectual property a shady disrepute that other kinds of property don't have. The ethical conundrum is reflected in the curiously indirect way that intellectual rights have been protected in most Western countries—until recently. Traditional copyright and patent laws in the United States, for example, don't identify ideas, inventions, and works of art as property but instead create a derivative product—a monopoly right to use the work—which then becomes property.[15] This practice may seem like splitting hairs, but it nicely evades the need to redefine a sacred thing as profane. Thus the musical performance we'll give you is a priceless work of art that humanity will cherish forever, but the crystal case in which we ship it is practical and will cost you $13.95. Another symptom of the problem is the blatantly emotional content of many arguments against intellectual property. A frequently quoted example is an excerpt of a letter that Thomas Jefferson wrote in 1813 to Isaac McPherson:

> He who receives an idea from me, receives instruction himself without lessening mine, as he who lights a taper at mine, receives light without darkening me. That ideas should freely spread from one to another over the globe, for the moral and mutual instruction of man, and improvement of his condition, seems to have been particularly and benevolently designed by nature, when she made them, like fire, expandable over all space, without lessening their density in any point, and like the air in which we breathe, move, and have our physical being, incapable of confinement or exclusive appropriation.

The word "moral" in this statement is key. While the technical message is that works of the mind aren't naturally scarce and so don't have intrinsic economic value the way a piece of land might, the actual point is that the fire of the mind is holy and not to be bought and sold.

The conceptual difficulties with intellectual property disappear at once, however, if you simply accept that it is not a natural right at all but a deceptive business practice. Jefferson himself understood this point and even wrote about it in a piece called "Thoughts on Lotteries":

> If we consider games of chance immoral, then every pursuit of human industry is immoral; for there is not a single one that is not subject to chance of some gain. In all these pursuits, you take some one thing against another which you hope to win. These, then, are games of chance. Yet so far from being immoral, they are indispensable to the existence of man, and everyone has a natural right to choose for his pursuit such one of them as he thinks most likely to furnish him subsistence.[16]

He was obviously groping for words when he wrote this piece, since the games in question are certainly immoral—and were even more so when he was writing. But he clearly intended to say that gamblers' attitudes are central to economic life and thus something you must reconcile with the religious foundations of the law, even if it requires radically rethinking some basic assumptions. His impish suggestion that we categorize these activities as moral wasn't serious and, of course, isn't what actually happened. Instead we just declared certain kinds of immoral activity to be legal. No doubt Jefferson would approve of this development if he were alive today and agree with the view that keeping ideas secret is, in many cases, a noble enterprise. But he would have a lot of trouble with criminalizing their theft.

The many people strongly in favor of reinforced intellectual property rights hate the idea that it's just business chicanery, and work hard when ethics comes up, to change the subject and allege that they're saving capitalism from socialism. That's obviously incorrect, since saving capitalism from socialism would mean getting government out of the poker game as much as possible, not using government to prejudice the outcome as much as possible in favor of yourself. Government involvement may be desirable, but saving capitalism from socialism is certainly not why.

It is unclear how much stock we should put in Jefferson's opinions about property rights. Notwithstanding his historic accomplishments as statesman, political scholar, and scientist, he was a notoriously poor businessman.[17] He left office in extreme debt and died insolvent. Moreover, he didn't free his slaves in his will, a decision his overseer, Edmund Bacon, attributed to the near bankruptcy of his estate.[18]

While at least one Internet gaming site says that Jefferson played poker, this claim is not very believable given his history of financial bungling.[19] If he did play, he must have lost a lot of money. But he definitely understood the importance of risk and was always open to learning important new things. It is reasonable to speculate that he might have eagerly agreed to lessons.

Everyone please ante up. Tom's in.

5

PATENTLY ABSURD

Patents are devilishly difficult to understand. The underlying theory is simple enough—granting inventors exploitation monopolies as incentives to invent—but there is some vexing fine print. For one thing, despite the billions of dollars that ride on patent decisions each year, it is difficult to think of any living inventor who is rich. Also, patent agencies and courts are notorious for incorrectly identifying who invented things. Legendary cases of injustice include the Armstrong regenerative circuit and frequency modulation patents, the Farnsworth television patents, and the Damadian magnetic resonance imaging patents, but there are countless others.[1] Instead of discovering who invented what, the courts effectively *define* who invented what through a series of findings, which then constrain all subsequent decisions through precedent. Not only do these decisions routinely misidentify inventors, they *create* inventors where none actually exist by granting patents for commonly known things.[2] Over time, this flawed discovery system generates a maze of fictions about who did what that requires years of study to master. People sometimes complain that it resembles a carnival or magic show, in which you watch dazzling illusions while the real action takes place invisibly backstage. Patent laws are confusing not because of their underlying theory or wording but because of the strange parallel universe of truth they generate.

In the past, these contradictions were tolerable because they didn't intrude much into everyday life. You would hear people say that, yes, it was a terrible shame that Major Armstrong failed to profit from his inventions, and an even greater shame that he committed suicide in despair, but look at the wonderful commercial radio industry we got! Nobody cared whether patent law protected actual inventors. They only cared whether it efficiently assigned monopoly rights to individuals capable of making enterprise, thus improving life for everyone else. But our recent transition to the Information Age has caused the logical loose ends of patent law to begin encroaching on intellectual matters that many people thought were off-limits and about which they care about very deeply—commonly held notions of right and wrong, for example, the sanctity of the human body, or the fact that two plus two equals four.

The encroachments of patent law are often cloaked in technical language. Here's an example. The U.S. Supreme Court recently made the preposterous ruling that the chemical processes of your body are not laws of nature. It did so in two steps. First, it ruled that laws of nature cannot be patented.[3] Then it ruled that gene sequences can be patented.[4] The inescapable conclusion is that gene sequences are not laws of nature. Here's another. The Court first ruled that algorithms or mathematical formulas, like laws of nature, cannot be patented.[5] Then it ruled that computer software can be patented.[6] Thus it ruled that software does not consist of algorithms or mathematical formulas! This revelation evokes howls of laughter from software engineers when they gratefully interrupt their late-night tasks of writing mathematical formulas to learn that "algorithm" is not a synonym for "computer program," as they had previously thought.

The source of these logical incongruities is not patent clerks or justices, who are, for the most part, motivated and well-meaning people, but legislators, who instruct them to grant patent rights liberally, follow traditions of legal precedent, and specifically *not* make decisions on the basis of scientific principles or common sense. The ostensible purpose of the last injunction is to guard against abuse. Not surprisingly, patents often violate both established scientific principles and common sense. For example, there continues to be a steady stream of awards for perpetual motion machines, even though they violate the second law of thermodynamics.[7] One of these perpetual motion proposals also invokes nonexistent principles of antigravity.[8] A patent was recently awarded for compressing data, no matter how random it is, without information loss, which violates the Shannon information theorem.[9] Then there is the long list of fun party patents such as motorized ice cream cones, pillows with retractable umbrellas, inclining coffins, methods for exercising cats, doggie poop freezers, centrifuge birth aids, and methods for swinging on swings.[10] These aren't a big concern, however.

The fictions of patent law are not, unfortunately, merely artificial monopoly boundaries but constraints on what we can say or do. The immediate reason is that lawsuits, even frivolous ones, can so damage a person financially through legal costs and lost work time that they can function as privately levied fines. A notorious instance of this problem is Uri Geller's series of lawsuits against stage magician James Randi, who alleged that Geller's bending of spoons with his psychic powers was fraud.[11] The suits were eventually dismissed, but Randi was so financially taxed by the legal costs that he needed donations to help defray them.[12] The same principle applies to patents. UCLA geneticist James Grody had to stop research on congenital deafness linked to the Connexion 26 gene because the owner of its patent, Athena Diagnostics, demanded a fee he could not pay.[13] Model railroader Bob Jacobsen received threatening letters and a bill for $203,000 from KAM Industries for alleged patent infringement after he wrote and published software that enabled a computer to control model trains.[14] Stephanie Cox, owner of a small shop called Pufferbelly Toys, was visited by agents of the Department of Homeland Security and politely asked to stop selling the Magic Cube, an item that allegedly infringed on the patents for Rubik's Cube—an allegation that later turned out to be false because Rubik's patent had expired.[15] Programmer Avery Lee had to stop giving away his open-source code VirtualDub because Microsoft alleged that its compatibility with their packaging protocols infringed on their patents.[16]

The specific targeting of individuals for patent infringement is relatively rare, but not because the law protects people from intimidation. It's just economics. The legal costs of suing someone are so high that the plaintiffs can't make money unless they target defendants with deep pockets. Cases like Avery Lee's are prosecuted at a loss as a defense against market penetration by competing businesses. Another important factor is that technical experts, although outraged, don't want to become victims like Randi.[17] They do, however, make jokes out of frustration. Blogger Joseph Palmer wrote:

> I think we need a jury system. A patent challenge would involve calling a panel who are "skilled in the art" expressed in the patent. The jury would be paid by the challenge fee, and provided with pizza and beer. Then the patent examiner would describe the problem to the jury, and give them three hours to see if they could come up with the same solution that was covered in the patent. If the examiner judges the solution to be similar enough to the patent, then the idea expressed in the patent would be judged invalid.[18]

While satire of this kind will probably be safe for a long time, criticism of specific judgments, especially if it's competent and clear, will most likely become more dangerous as computer technology brings down the cost of suing someone.

At the corporate level, by contrast, lawsuit predation is serious. The University of California and a company called Eolas recently settled a $521 million judgment against Microsoft over infringement of Internet browser protocol patents.[19] Research in Motion, the owner of the popular Blackberry wireless service, paid a settlement of $450 million for alleged infringement of wireless protocol patents.[20] Rambus's $306.5 million judgment (later reduced to $133.5 million) against Hynix for infringement of Synchronous Dynamic RAM patents just survived appeal.[21] Toshiba paid a settlement of $288 million to Lexar for alleged infringement of Flash memory patents.[22] City of Hope Medical Center won a $500 million judgment against Genentech for alleged licensing violations regarding a key recombinant DNA patent.[23] Amazon.com just settled a patent infringement suit with IBM that threatened to involve "hundreds of millions" of dollars.[24] The enormous amounts of money involved in these suits reveal a constraint on what we can say and do that is more subtle, and probably more important in the long run, than threat of legal action against individuals: the threat of legal action against an individual's *employer*. Just imagine, for example, what would happen to an employee of Microsoft or IBM who revealed a fact that was true but nonetheless detrimental to the company's patent position. There's an analogous constraint on tenured university scientists, although it's more subtle. A university researcher who published findings detrimental to a big donor or that infringed on the patents of an aggressive company or competing university wouldn't face dismissal, but he or she would certainly face long waits outside the President's office and endless grief with appointments, promotions, fellowships, lab space, and other things professors care about. More importantly, there would be punishment by government funding agencies, who are charged with spending money in ways that will benefit the economy and who don't want to be sued. This kind of intimidation can be effective even at very high levels. Microsoft, for example, threatened to sue several Asian governments if they used Linux.[25]

The restriction on our behavior implicit in patent practices is not as temporary as proponents often allege. In theory, patent monopolies are granted for a limited time (now twenty years) and then expire, but they can, and do, get extended.[26] One way is simply for the legislature to pass a law granting extension to a specific patent. There are precedents for this kind of extension, although they are rare. Under U.S. law, patents can also be extended in compensation for delays in the patent office or, in the case of drugs, delays in government approval for use. But the extension method that's most important is modifying the original invention in some way and then re-patenting the modified version as "new." For example, GlaxoSmithKline successfully re-patented the antibiotic cefuroxime by arguing that cefuroxime axetil is a different drug from cefuroxime sodium, even though the active part of the molecule in both cases is the same.[27] A very large fraction of software patents are also in this category.

The indirect financial constraint on a person's actions also explains why university investigations so often appear abstruse and economically irrelevant. The scientists know perfectly well what things are most important economically, but they are prevented from working on them by their sponsors' vulnerability to lawsuits. Far from enjoying this state of affairs, they find it maddening, especially since the obstacles they're avoiding are legal fictions, the exact opposite of science.

The capriciousness of patent law and the very high stakes involved have traditionally encouraged organizations to patent as aggressively, rapidly, and broadly as possible, even at the risk of generating patents that seem a bit ridiculous.[28] People have understood since the early days of steam that there's no penalty for being wrong in a patent application, only for being late.

Moreover, the ultimate value of a patent is difficult to predict, since today's worthless claim can be tomorrow's gold mine if it can be construed as part of someone else's later invention. This is true even if the other inventor didn't know of its existence! Stanford University's patent for the fiber optic amplifier, for instance, lay dormant for years until David Payne independently thought of doping glass fiber with erbium atoms to make it amplify.[29] Payne's invention worked so well that it was adopted as a standard in the optical communications industry. That made the previous patent valuable, so Stanford and its partner Litton brought suit and obtained compensation.

The practice of patenting first and asking questions later has now taken an alarming new turn. The rush to patent everything in sight, notably in software, is increasingly stigmatizing the acquisition and mastery of knowledge that, until recently, was in the public domain. For example, in early 2003, SBC threatened to enforce patents covering the use of frames on Web sites, a generic and widespread computer programming practice.[30] In other words, SBC would sue you if you typed certain things and then showed other people what you'd typed. Acacia Technologies owns several broad patents on the use of streaming media (Internet delivery of movies), another industry standard technique, that it has used successfully against adult Web sites, universities, and large media companies.[31] This means that Acacia might sue you for taking digital home movies, processing the files with computer programs that you write yourself, and putting the processed files on the Internet. Cisco owns a broad patent on the entire idea of combining voice, video, and data delivery to the home through the Internet that is expected to cause endless trouble.[32] Microsoft, which has set itself the goal of obtaining 2,000 to 3,000 new patents per year, is attempting to patent such astonishing things as verb conjugation and a method for distinguishing "good" Internet behavior from "bad."[33] Cingular has applied to patent smileys.[34]

The source of this troubling trend is the recent strengthening of patent laws, notably instructions issued by legislatures to grant patents for "everything under the sun."[35] This instruction is followed by the words "made by man," but the courts have now decided that this phrase means "understood by man," which makes it a fairly insignificant constraint. As a result, the Alice-in-Wonderland world of patent law has expanded so aggressively that it seems to have no limit. Once the courts have ruled that computer programs aren't algorithms and gene sequences aren't laws of nature, there doesn't seem to be much preventing us from having patents for wind, dirt, or the act of thinking. In extreme moments of cynicism you can imagine airplane manufacturers and airlines paying licensing fees for the use of oxygen, or farmers and tractor manufacturers paying fees to plow. You can even imagine Asians and continental Europeans being charged for using English, on the grounds that someone else thought of it first.

While patenting the wind is still far off, outrageous patents that send technical experts into apoplectic screaming fits are increasingly commonplace. An example is Amazon.com's patent on one-click purchasing, with which it successfully sued Barnes and Noble.[36] Another is Blackboard's recent claim on the entire concept of online education.[37] Yet another is the patenting of embryonic stem cells.[38] Utter revulsion is also routinely expressed toward outrageous patent *strategies,* such as allowing applications to remain in process in the patent office, and thus secret, for many years, like submarines hiding under water. The patent then gets granted, allowing its owner to torpedo competition that has been using the invention unknowingly for years. Famous submarine patent attacks include the University of California's $100 million suit against Monsanto over the recombinant bovine

growth hormone Posilac and Chiron's suit against
Genentech over the cancer drug Herceptin.[39]

Thus today's complaints against patent practices differ from the grumbles of the past not just in sheer numbers but also in tone. Many articulate, well-educated
people have begun to express an exasperation so extreme
that it amounts to disrespect. Technical people often
have difficulty articulating exactly what troubles them,
so their complaints often come out as entertaining
tirades posted anonymously on blogs. But the problem
is straightforward. Newly aggressive patent practices are
increasingly violating a principle that has been with us
since Roman times and is built into our societies at
many levels, including our religions: the laws of man
flow from the laws of nature and are subservient to
them. Patenting nature is transparently immoral. So is
patenting reason, since reason and nature are one and
the same. Thus the current problem with patent law is
more serious than the bellyaching of a few jaded engineers. It's a crisis of legitimacy.

This unhappy state of affairs has now led to loud
cries for tort reform, both from concerned individuals and from businesses being bled dry by legal
costs.[40] Yet while legislatures may enact reforms soon,
most people think they are unlikely to do what it
takes to restore respect for technical law: ban patents

for methods of reasoning, methods of communicating, discovery of things widely viewed as self-evident, and discovery of phenomena that occur on their own when humans are not present. The economic cost would be too great. For better or worse, our society seems committed to making it easier to own technical knowledge, even if it means alienating rank-and-file technologists and abandoning science for fiction.

Meanwhile the legitimacy crisis continues to metastasize. Young people engage in massive civil disobedience by copying music, movies, and software to each other. Software companies persecute independent inventors, the people that patent law ostensibly protects, as dangerous guerrilla warriors.[41] Bloggers give up in disgust, leave technical life, and turn to more profitable pursuits such as real estate sales, bond pricing, medicine, and patent law.

If one is forced to think realistically about the situation, it's difficult to be optimistic.

6

THE NUCLEAR PRECEDENT

It's generally agreed that nuclear matters occupy a special category all their own. Nuclear reactions, particles, and so forth aren't part of our everyday experience, so they have a kind of abstract, otherworldly image, like demons, ghosts, or mysterious diseases. We speak matter-of-factly about the nuclear genie but wouldn't imagine speaking about the electric genie, the jet airliner genie, or the offshore oil platform genie.[1] We find it reasonable to ban nuclear reactors but find it unreasonable to ban parking lots, steel mills, or fertilizer plants.[2] Nuclear-free zones make sense to us in a way that fungus-free zones or weather-free zones don't.

This perception is only half right. While it's true that nobody knew about nuclear radiation until about one hundred years ago, when suitable detectors were invented, it's also true that radiation was always there at low levels and always mattered. The mutations in our genes that it slowly induced presumably played a role in making us who we are. Much of this environmental radiation comes from the sky in the form of cosmic ray impact detritus.[3] There's a constant rain of about one muon every second per square centimeter at sea level, more at high elevations. Muons are very fast particles that pass completely through the body and so do a relatively large amount of damage. There's also the radiation you get from the decay of 4,000 or so radioactive potassium atoms in your body every second.[4] If you live near masses of granite, which are naturally radioactive, your exposure is considerably greater. Then there are the 1,200 decays of radioactive carbon in your body per second, about 50 of which occur in your DNA. And so on. The sum of this environmental radiation isn't great by health measures, but it isn't zero either. There is literally no such thing as a nuclear-free zone on Earth.

Nonetheless, our feeling that nuclear energy is somehow man-made and expendable has made it easier for us to shun it, thus creating a powerful precedent for knowledge sequestration. Most people find it perfectly reasonable that responsible government agencies should keep this fearsome thing locked up and out of the hands of mischief makers. Most people agree that individuals who want to know about this stuff are precisely the ones who shouldn't be allowed to do so. Most people are at peace with the notion that "unauthorized" acquisition of this knowledge should be a crime, that is, a transgression against the state punishable by fine, imprisonment, or death.

But there are a few logical loose ends. For one thing, exactly who should do the authorizing is somewhat unclear, since we want the knowledge to be invisible to everyone, including ourselves. As a practical matter, it's a regulating agency, which then effectively "owns" the knowledge and has the power to bestow learning rights, in the way a landowner might bestow access. Then there's the problem of how these crimes should be prosecuted, tried, and punished. A conventional court proceeding is problematic because it requires exposing to public scrutiny the very knowledge you want to keep secret. Secret trial is a possibility, but it's unconscionable and also illegal in advanced countries. There's also the possibility that people inside the agency privy to the knowledge might do things with it under the veil of secrecy that you don't like, such as plan nuclear power plants in your town. There's also the possibility that they might simply lie about how important their secrets are and how effective their security has been.

The enforcement conundrum is not merely theoretical. In a 2002 letter to the Speaker of the U.S. House of Representatives, Attorney General John Ashcroft discussed the issue of unauthorized disclosure:

> Clearly, that only a *single* non-espionage case of an unauthorized disclosure of classified information has been prosecuted in over 50 years provides compelling justification that fundamental improvements are necessary and we must entertain new approaches to deter, identify and punish those who engage in the practice of unauthorized disclosure of classified information.[5]

Since it is not possible that only a *single* non-espionage violation of the Atomic Energy Act occurred over fifty years, the obvious implication is that the criminal provisions of the act not justified on grounds of espionage are difficult to enforce.

This conclusion is corroborated by the history of three famous public nuclear security cases—the *Progressive* article of Howard Morland, the atomic bomb thesis of Princeton undergraduate John Aristotle Phillips, and the Los Alamos downloads of Wen Ho Lee.[6] In each of these instances, highly publicized and potentially grave classifi-cation breaches were, in the end, neither confirmed nor denied by the government, and the constitutionality of the Atomic Energy Act's criminal provisions wasn't openly challenged in the courts.

The *Progressive* case was the first test of U.S. nuclear secrecy laws in open court.[7] It concerned a failed at-tempt by the government in 1979 to stop publication of an article by freelance journalist Howard Morland about hydrogen bomb construction.[8] The government sought and obtained a temporary injunction against publication of Morland's article on the ground that his "theories" about nuclear weapon design constituted unauthorized disclosure of restricted data prohibited by the Atomic Energy Act. It argued that this was true regardless of whether someone had illegally disclosed things to Mr. Morland or he had figured them out on his own from information in the public domain, as he claimed to have done. The presiding judge, Robert W. Warren, concurred and issued the injunction. But dur-ing the subsequent appeal process the government simply dropped the suit. The reason it gave was that the matter had become moot on account of the leak in the interim of a letter written by weapons hobbyist Chuck Hansen to U.S. Senator Charles Percy, in which similar ideas were proposed.[9] After Morland's article was published, however, it became clear that Hansen's ideas were fundamentally different from Morland's.[10] A more believable reason is that the government feared it might lose the appeal.

One of the things the government was attempting to enforce in the Morland case was its need to censor *thinking* in the interest of national security. Judge Warren felt that this point was so important that he included a passage about it in his opinion:

> The government argues that its national security interests also permit it to impress classification and censorship upon information originating in the public domain, if when drawn together, synthesized and collated, such information *acquires* the character of presenting immediate, direct and irreparable harm to the interests of the United States.[11]

He went on to observe that it wasn't just the publication of numbers that was dangerous but also the exposition of certain concepts—things that would save competing states money and time wasted going down blind alleys. In other words, the "data" restricted by the Atomic Energy Act included ideas. Judge Warren then sided fully with the government and ruled, based on evidence supplied in secret to the court, that the Morland article constituted not only a violation of the Atomic Energy Act but also a threat to national security so great that it could also be censored under other statutes. He likened its publication to disclosing troop movements in time of war and wrote that the threat to life it presented manifestly overrode freedom of speech, even though "any prior restraint on publication comes into the court under a heavy presumption against its constitutional validity." No sensible jurist, he contended, would place freedom of speech above the right to live.

The government didn't have at its disposal the kinds of punishments required to prevent ideas from propagating—destruction of the creator's livelihood, execution of his children, grisly public torture, and so forth—so it failed to stop publication of Morland's article. This does not imply that its concerns were unjustified or that the article was innocuous. Indeed it suggests the opposite. The government couldn't fine or imprison the perpetrators after the article was published because a public trial would have revealed which speculations were the dangerous ones, thus further compromising national security. What it does imply is that keeping concepts secret over the long term is extremely difficult and perhaps impossible.

The other notorious case of illegal logical synthesis was the 1977 Phillips bomb design.[12] Under the supervision of Professor Freeman Dyson, who claimed not to have helped, Princeton undergraduate John Phillips wrote a dissertation about how to build an atomic bomb. His sources were library books and unclassified documents that he obtained for free from the government. He brought the information in these sources together logically, supplemented it with some thinking and computation of his own, and came up with a weapon design that he thought would work. He wrote up this design as his thesis and also built a mockup, which he kept in his dorm room. The story became celebrated in the press, at which point the FBI confiscated both the thesis and the mockup, their grounds being that Phillips, by merely thinking about nuclear weapons, had violated the Atomic Energy Act. The government did not press charges.

The Phillips case had the interesting side effect of revealing the severity of the conflict between national security needs and the educational objectives of universities. Phillips was not a revolutionary bent on destroying nations but a rather average student who played a cowbell in the marching band and wore a tiger suit at Princeton athletic events. He did the research on his own because he was interested, and went on after graduation to run a software company. The danger to the state he represented was not as a mad scientist or spy but simply as a bright young person asking important questions on his own. Unfortunately, universities were invented precisely to encourage such behavior. Similar events have happened routinely in other universities around the world, albeit not so visibly. The threat is thus from the university environment itself, not just one or two renegade students. Academic researchers often feel thrilled at the idea of being considered dangerous, if only for fifteen minutes, but they shouldn't. Fear of security consequences prevents many of their most important ideas from being funded.

The security danger posed by academic attitudes toward scientific inquiry is also behind the recent spate of security scandals at U.S. nuclear weapons laboratories.[13] The undesirable behavior is sometimes described as "cowboy culture," but it has nothing to do with cowboys and everything to do with professors, in particular, the academic practice of thinking independently about what is and is not scientifically correct.[14] This skill is essential for making new discoveries and, more importantly, for making new engineering designs that work. You don't want to fly in an airplane designed by committees of obedient workers. You want to fly in an airplane designed by cowboys. But after the design has matured, your concern shifts from making things work to preventing your competition from making things work. Cowboys then become potential security liabilities, so you get rid of them or give them something harmless to do. There was a great joke floating around both the Los Alamos National Laboratory and the Lawrence Livermore National Laboratory during the first Bush administration, when a barrage of seemingly meaningless health and safety directives came down from Secretary of Energy James Watkins. Admiral Watkins, the story went, had no idea what people should do, so he fell back on his military training and ordered them to police the area.

The case of Wen Ho Lee at Los Alamos, however, was less a cowboy attitude problem than a deadly serious espionage problem.[15] The facts of the case are so confused by strategic press leaks, legal defense tactics, and prosecution mistakes that the public may never know the exact nature of Mr. Lee's transgressions. Senator Arlen Specter, a member of the Senate Judiciary Committee, read the following statement into the U.S. Congressional Record on September 20, 2001:

> One great tragedy of the Wen Ho Lee case is that the entire truth will likely never be known. As a consequence of an inept investigation, the government has lost the credibility to claim that its version of events is the absolute truth. Dr. Lee also lacks the credibility to

tell the definitive tale in this case: he repeatedly lied to investigators, created his own personal nuclear weapons library without proper authority, copied nuclear secrets to an unclassified computer system accessible from the Internet, and passed up several opportunities to turn his tape collection over to the government.[16]

Dr. Lee did something wrong, but precisely what it was couldn't be determined with the certainty required to convict him of a crime in open court. He eventually pleaded guilty to one felony count of mishandling classified information as part of a plea bargain.[17] His lawyers released the following statement on his behalf on September 11, 2000:

> On a date certain in 1994, I used an unsecure computer in T-Division to download a document or writing relating to the national defense (File 14) onto Tape L. I knew at the time that my possession of Tape L outside the X-Division perimeter was unauthorized and that, under Los Alamos National Laboratory Directives, I was not permitted to have Tape L outside the X-Division perimeter. I retained Tape L and did not deliver it to an officer or employee of the United States entitled to receive Tape L.[18]

This occurred after he was denied bail and held for 267 days in solitary confinement.[19]

While Lee's admitted wrongdoing was considerably less than the fifty-nine crimes listed in his original indictment, it was nonetheless significant, because his motives were never satisfactorily explained. His attorneys reported that he had made the downloads and copies of sensitive files in anticipation of being laid off and seeking work at other laboratories, perhaps even ones abroad. However, everyone in a nuclear weapons laboratory knows that you don't do that. Even in 1994, all Los Alamos employees were required to attend regular refresher briefings in which security officers repeatedly emphasized that you don't download or otherwise questionably handle files marked "Protect as Restricted Data" (PARD). There were always lots of jokes about PARD after these briefings because it's an amusing word that you can rhyme with many things while venting frustration at having to sit through the briefings.

Unfortunately, this kind of circumstantial evidence can't be ignored if you want to stop knowledge leaks. Copying and transmitting information leaves few traces, especially if it's done by computer. The damage also accumulates over time, like damage from a leak in your roof, so that over the long run, catching only some spies is functionally equivalent to catching none of them.

The Lee incident thus revealed a profound conflict between national security needs and civilized rules of criminal evidence, specifically, treating the accused as innocent until proved guilty. You can't have both. It's scarcely surprising that government officials, faced with the need to prevent massive loss of life and with a legal system constructed before such a need existed, sometimes indulge in what appear to be unethical and extralegal practices—such as using technicalities to keep an accused spy in solitary confinement for nine months and then "apologizing" to him after he is found innocent of 98 percent of the charges.

The same conflict also occurs in Europe, of course, although it plays out more subtly because of international complications. Thus engineer Gotthard Lerch is presently undergoing a lengthy and complex trial in Mannheim, Germany, for violating "export controls" in connection with A. Q. Kahn's attempts to assist Libya in building atomic weapons.[20] The defense has complained about the poisoning of public opinion against the defendant, just as occurred in Wen Ho Lee's trial, and the prosecution encountered further difficulties when asked to explain in public precisely what technology couldn't be sold to the Libyans and why it constituted a threat to European security. The trial, which began in March 2006, was halted in July 2007. A new trial is expected to begin in 2008.[21]

These and other incidents demonstrate rather starkly that modern civilization rests on two mutually exclusive kinds of thinking—one embodied in the free speech guarantees in the First Amendment of the U.S. Constitution, the other in the Atomic Energy Act.[22] The latter is not exclusively nuclear, of course, and comes to us through a series of court decisions about wartime security.[23] However, the fact that a decisive court battle over the constitutionality of the Atomic Energy Act has been successfully deferred for over fifty years reveals that neither principle has the political support to become preeminent. Instead, each has a domain of validity in which the other makes no sense. Banning thinking is a ludicrous concept in a newspaper editorial office, but so is promoting free speech in a nuclear weapons laboratory. As a practical matter, we are a strange chimeric hybrid of Athens and Sparta.

Notwithstanding the few highly publicized human rights "triumphs" like Morland's and Wen Ho Lee's, nuclear realities have pushed us decisively in the direction of Sparta. It seems so obvious to people that even *thinking* about nuclear matters should be banished that they can't see the obvious generalization to banishing other kinds of thinking. But the unhappy truth is that nuclear knowledge is not uniquely dangerous. Chemical and biological warfare, robot technology, new kinds of missiles, computer attacks on the financial system—all have the capacity to generate massive loss of life equal to or greater than the threat posed by nuclear weapons.[24] Officials responsible for public safety understand this, and so they constantly fret about restricting knowledge relevant to all of these technologies. Our experience with the Atomic Energy Act reveals that this can be done without statutes that directly confront human rights guarantees in the courts. But these indirect secrecy practices don't stop determined foreign states from acquiring the knowledge through espionage, and so don't stop the long-term threat of state-sponsored terrorism. That implies that human rights guarantees, over time, threaten security. Weakening the First Amendment right of free speech is politically unthinkable in the United States at the moment, but another successful terrorist attack on New York might change a lot of minds.[25] Indeed, no European country, even Great Britain, has such a right, presumably for this very reason. We talk nonchalantly about our having opened a Pandora's box of technical knowledge, but the other Pandora's box we've opened is that of knowledge sequestration.

Unfortunately, the idea that knowledge not properly controlled is bad transfers easily from security matters to economic ones. Economic competition is, after all, a kind of "war" in which the losing side pays a steep price, perhaps even an unacceptable one. While there's no direct link between the Atomic Energy Act and nonmilitary intellectual regulatory legislation such as the Digital Millennium Copyright Act, the conceptual similarities are unmistakable, as are the similarities in their human rights implications.

Thus our apparently innocent decision to treat nuclear knowledge as different from other kinds of knowledge wasn't so innocent. It set a precedent that has now led, by small steps, to a significant and growing threat to our freedom to reason and learn.

7

THE FACTS OF LIFE

It's hard to imagine anything more wonderful or natural than learning about life. Not only are living things beautiful and interesting, they're part of our milieu—rather, we're part of theirs, since the ecosphere could survive nicely without people but not vice versa. Conodonts, trilobites, and carboniferous forests all came and went before humans were even a glimmer in their Maker's eye, as did the dinosaurs.[1] Humans might have wiped out the mammalian megafauna, the thylacine, and the dodo, but they certainly didn't create these animals.[2] Moreover, life processes are extremely important to us economically. Living things feed, clothe, and house us. They create our environment.

They generate the air we breathe. They threaten us with physical harm, predation, and disease. They regulate our bodies and grant us time on Earth—or not. It makes sense that we should learn as much about them as we possibly can.

This conclusion is overly simplistic, of course. Knowledge about life, like any other kind of knowledge, can be excessive. Everyday versions of this problem make good jokes. Discussing how cats eat their breakfast isn't polite. People who grew up on dairy farms often don't care for dairy products. Students who pay too much attention in human reproduction, the best course in high school for everyone else, become tiresome. The more serious versions are anything but joking matters. A person intimately familiar with diseases might intentionally infect a population of humans or agricultural plants and animals, thereby causing an epidemic.[3] A person thoroughly educated in gene manipulation might create dangerous forms of life, such as giant hermaphroditic lizards, virulent new diseases, or people specially designed to be slaves or soldiers.[4] A person knowledgeable in the workings of cell machinery might use microsurgery to usurp normal embryonic development and create monsters for amusement or for spare parts.[5] Clearly, you don't want just anybody learning how to do these things.

There are, of course, also commercial reasons for keeping key facts of life out of the public domain. For one thing, it raises research costs for competitors trying to copy or improve your proprietary agricultural or medical products. That's one of the reasons not just anybody can make transgenic insect-resistant soybeans, cotton, and corn,[6] drought-resistant wheat, rice, and tobacco,[7] or salt-resistant rapeseed,[8] even though it's supposedly easy to do. A murky knowledge base also allows you to make risky technological investments—heart-friendly pigs,[9] rapidly growing salmon,[10] medicinal chickens,[11] and so forth—with reasonable expectation of profit. It has the further salubrious effect of raising costs for people wishing to criticize your products on scientific grounds—for example, that they inadvertently poison people,[12] pollute the wild gene pool,[13] or inflict collateral damage on beneficial insect populations.[14]

Unfortunately, the seemingly commonsense conclusion that some biological knowledge must be kept out of the public domain has enormous implications. To achieve this result not only must you censor dissemination of certain findings, you must also discourage their subsequent discovery by someone whom you cannot censor! This can't be done without interrupting the intellectual traditions of the science itself—especially in a discipline with only modest laboratory equipment needs. You must eradicate not only ideas but the seeds of those ideas and the process of synthesizing them logically into a larger whole. This is like battling a weed infestation or a cancer. This dilemma is, of course, exactly the one faced in curtailing the spread of nuclear knowledge—an analogy that makes many professional biologists understandably unhappy. However, the facts speak for themselves. Not surprisingly, the solution now being implemented in biology is similar to that already implemented in physics—as are the shortcomings of that solution. There is no biological analogue of the Atomic Energy Act in any country at the moment, but this is quite irrelevant because experience with the act over time revealed that its criminal provisions were not effective at stopping the spread of knowledge. The truly effective measure has been "voluntary" self-censorship enforced with threats of firing, loss of grant support, and public ostracism for willful research interests in the "wrong" directions or public speech about subjects everyone knows to be taboo. That's occurring now in biology.[15] The central problem with this practice, however, is that it stops *your* progress and that of would-be terrorists but not that of determined competition with adequate financial resources.[16]

Ironically, the most immediate threat posed by biological knowledge does not involve new high-tech discoveries at all but simply abuse of well-known and very ancient diseases. Many of these diseases are unfamiliar to the public, perhaps to many physicians, because they're now controlled so effectively. They're nonetheless out there waiting to be exploited for evil purposes.[17] Commonly cited examples include plague, cholera, typhoid, tularemia, anthrax, Q fever, trench fever, yellow fever, dengue fever, encephalitis, Marburg virus, and smallpox. There is an equally long list of potential threats to agricultural plants and animals. The only technological issue involved with turning these diseases into weapons is figuring out how to infect many victims simultaneously, thus initiating an epidemic that overwhelms the health management system. Fortunately, this is not a trivial matter—although several countries made major advances in weaponization during the Cold War.[18] Such research presumably continues in secret today, despite being banned by treaty.

The nature of the weaponization problem is best explained with a concrete example. If you fly economy class a great distance on a crowded airplane you are likely to catch cold. The airline would rather that you didn't catch cold, but unfortunately it has no power to protect you. The few infected passengers who have irresponsibly chosen to fly with you enter the plane undetected and, once aboard, touch things, sneeze into the air, and so forth. Some of them are in the early stages of their cold and so don't even know that they're contagious. Thus you can just imagine what would happen if some unscrupulous people boarded the plane intending to infect you *on purpose* with a serious disease like smallpox.[19] They might do this suicidally, by coming aboard infected and coughing on everybody, or by inoculating themselves first and then bringing aboard a little bottle of infectious material and dusting it all over the lavatories, galleys, air intakes, and elsewhere. The result in either case would be hundreds of very sick people, each of whom would infect many more, who then would infect many more before the nature of the problem was discovered and proper countermeasures were taken. The toll from such an attack would be several thousand deaths *per plane.* Weaponization in this case just means making the dust.

A scenario much like this was the subject of the chilling "Dark Winter" war game conducted in the summer of 2001—ironically just three months before the 9/11 attacks.[20] While the purpose of this game was to educate people responsible for bioterrorism preparedness about the nature of the problem they faced, it did considerably more than that. It convinced many people that such an attack simply must not happen. Rumor has it that former Senator Sam Nunn, who played the U.S. president in the game, stopped play after two days because he had "seen enough." The official transcript reports only that the game lasted two days and covered a simulated period of two weeks. The premise was that terrorists had released weaponized smallpox in three shopping malls—in Oklahoma City, Philadelphia, and Atlanta—initially infecting 3,000 people. About 30 grams of powder would have been sufficient. The simulation began one week (the incubation period) after the hypothetical attack with reports of 12 confirmed cases of smallpox in Oklahoma, 14 suspected cases there, and rumors of cases in Georgia and Pennsylvania. It ended with an unfolding global catastrophe: The death toll was 1,000 and climbing. There were 16,000 new infections, 14,000 of which had developed within the previous 48 hours. The disease had spread to twenty-five states and ten countries. Hospitals were being overwhelmed. Food deliveries were failing. Businesses were closing. Highways were jammed with people fleeing danger. Estimates were that one month hence, 1 million people would be dead and another 3 million infected, with no end in sight. Plans were being drawn up for retaliating against the perpetrators—whoever they were—with nuclear weapons.

In light of the "Dark Winter" threat, it's scarcely surprising that many people oppose allowing *any* public access to fundamental knowledge about dangerous diseases, notably and particularly any scientifically correct understanding of how they work.[21] The screams of protest became particularly shrill around the time of the 9/11 attacks, as extensive genomic details about plague, anthrax, smallpox, and Q fever were published in relatively quick succession.[22] Proponents of open publication countered that censorship was inimical to science and that this particular knowledge was central to modern pharmacology, particularly to antiviral drug development, without which the AIDS epidemic would still be out of control.[23] But it was too late. Earlier that year, researchers in Australia had announced their discovery that a small genetic modification of the mouse version of smallpox made it 100 percent lethal.[24] It is now widely acknowledged, even among academics, that this information was extremely dangerous and perhaps should not have been published. The machinery of self-censorship began to be put in place at that time. It will be interesting to see how much fundamental knowledge on this subject remains in the public domain ten years from now.

Ironically, pathological microbe genomes aren't actually that dangerous.[25] The principles relating gene sequence to protein function aren't known, so sequences are useful mainly as guides to large, well-funded research efforts, not as blueprints to casual terrorists. Also, while you can indeed make diseases starting from computer files, as was recently done with poliovirus, only a very dumb terrorist would do so.[26] It would cost too much money. He would instead just get the virus from a stockpile or from someone with a polio infection.

The earnestness of the protests thus betrays a deeper, more troublesome problem that has little to do with either disease or warfare. It's the fear of ideas—the important kind that empower people and thus have consequences that are difficult to accurately predict. You sometimes hear this fear expressed as the concern that knowledge about life presents such potential for mischief that *nobody* should have access to it—especially scientists. You hear that the knowledge we already have is excessive and should not have been created in the first place. Many people feel that we should lock it away, throw away the key, and then prevent any further malfeasance by zeroing out public funding for people who make the "wrong kind" of discovery—or are likely to do so in the future. Some even feel that we should criminalize certain lines of inquiry to make sure that the unwanted knowledge isn't generated with private resources. But these arguments are misleading, for the fear being expressed is not actually about biology at all but about fundamental discovery, something that right now is occurring with especially great frequency in the life sciences.[27]

The unarticulated concern that reason itself might be unethical accounts for the extreme reaction many people have had to research on stem cells and cloning,[28] developments that, unlike the sequencing of microbe genomes, are extremely important scientifically.[29] While bans on such research have now been repealed in many legislatures around the world,[30] the hostility toward the knowledge itself, which caused the bans in the first place, hasn't disappeared.[31] The discussions now occurring are not about what humanity must know about life but about curing cancer, repairing severed spinal cords, forestalling senility, and maintaining competitiveness in health care enterprises.[32] New kinds of agricultural products are not on the table, nor is the creation of new kinds of pets or the mastering of morphogenesis, the principles by which genes encode the shapes and functions of living things. The implicit understanding is that public money is to be used for curing disease and prolonging life, *not* for understanding how life works and revealing this understanding to everybody else. Almost without exception, legislators would prefer that scientists improve health technology without generating troublesome new knowledge. Almost without exception, that's the message coming down to researchers from their funding agents.

The first cloned higher organism, Dolly the sheep, was important for reasons people often don't fully appreciate.[33] It certainly wasn't because her creators had invented a technology for reproducing higher organisms asexually. That aspect was fun to discuss, especially on the floor of the U.S. Senate, and made it terrific press as we got, in quick succession, CC the cat (Copy Cat), Ralph the rat, Idaho Gem the mule, Millie the cow, Prometea the horse, Dewy the deer, Snuppy the dog, and little piglets Noel, Angel, Star, Joy, and Mary.[34] It is very likely that we also got Hal (or Heather) the human, since cloning people isn't conceptually more difficult than cloning animals. There was, however, no public announcement of a cloned human, presumably because of the storm of public outrage that would have ensued. The one case of preliminary human cloning that became public was later exposed as a fraud.[35]

The importance of cloning lies in proving, once and for all, that the potential to become a complete organism resides in every cell in the body, not just in germ cells, and that the "switches" determining the function of a cell are volatile and can be reset. In the case of Dolly and other clones like her, the resetting was accomplished by slowing down the cell's metabolism through carefully managed starvation, which is rather like scrambling the memory of your computer by turning down the voltage of its power supply.

The issue at stake with cloning is considerably more than just philosophical musings about whether or not life begins at conception. It's the possibility of regenerating lost organs or limbs, the way a salamander's leg grows back when amputated, by correctly resetting the switches of a victim's own cells.[36] At the focus of most regeneration research are stem cells. These are simply cells in which some of the switches have been thrown but others have not, making them plastic in ways that the body finds useful, but not in ways that it doesn't. The most versatile stem cells are obtained from embryos, but others are found in every adult body and are involved in causing wounds to heal when you are injured.

Whether a disciplined focus by researchers on medical practicalities will quiet the critics of cloning remains to be seen. Discoveries about life as profound and shocking as Dolly are bound to occur occasionally, even if institutions do their very best to prevent them. You can imagine what would happen, for example, if somebody discovered a mammalian gene analogous to Antennapedia in Drosophila and performed recombinant experiments to demonstrate its function.[37] Modification of this gene can either cause legs to grow out of the fly's head where the antennae should be, or antennae to grow out of the fly's thorax where its legs should be. In either case, the result is a monster based on a throwback to the body plan of a more primitive ancestor. Our likely response to such a thing in a domestic animal would be to excoriate the discovery as reprehensible, marginalize the discoverers, criminalize the relevant "technology," organize United Nations condemnations, label the subject taboo, ban discussion from classrooms, and do our best to forget that the whole thing ever happened. Far from getting a trip to Stockholm, the people responsible would be hated, shunned, and vilified as terrorists.

A mild, but telling, case of this problem is now unfolding in Britain, where the government has moved to block research on certain human-animal chimeras—grafts of human cells with those of another species.[38]

Chimeras have been created and used in the laboratory for years and are particularly useful in organ transplant and tissue regeneration studies.[39] They have included such whimsical creatures as the goat-sheep chimera (geep), the sheep-human chimera, the turtle-chicken chimera, and the chicken with a quail brain.[40] Many people with heart conditions are themselves chimeras in that they have received replacement valves from pigs or cows. These valves coexist with their recipient's own tissues for many years. But the proposed creations pose a special ethical dilemma because they involve transferring a human somatic cell nucleus into an animal egg—bypassing the need for a human egg in making a clone.[41] This renders moot the arguments that cloning should be banned because it destroys the original nucleus of a human egg. It also sheds light on whether the "switches" of a cell reside in the nucleus or the cytosol and on how unique to a given species the switching signals are. Unfortunately, the only way we presently can detect these signals is to see how stem cells develop in a given milieu. Moreover, if the experiment succeeded, it would prove that the human egg contributed nothing other than mitochondrial DNA to the making of a person and thus that the full capacity to create life exists in every cell of the body, not just in its germ cells. That, in turn, would push people toward accepting the clear implication that our concepts about what it means to be human, and our laws based on them, need to be revised.

Troubling new ideas have always met terrible resistance. Most scientists will tell you from sad experience that this resistance is the norm, not the exception. Socrates was executed for promoting views that, while perhaps valid, constituted a threat to the state. Galileo was tried, convicted, and held permanently under house arrest for similar reasons. Opinion about birth control pills was so polarized in the 1960s that it required a U.S. Supreme Court decision to make them legal in all fifty states.[42] The theory of evolution is still widely despised and feared for reasons little different from those articulated in the Huxley-Wilberforce debate 150 years ago.[43] What's new in the case of cloning is the association, albeit remote, with threats to physical security, not only to ourselves but to our livelihoods, our environment, and our progeny. In many ways this development is the unhappy flowering of a seed planted inadvertently in the 1950s, when we decided that nuclear knowledge was sufficiently dangerous that it could be isolated from the body of "legitimate" knowledge and banned. It is a cruel twist of fate that the same reasoning is now applied to the fundamental understanding of life.

8

CLONE WARS

Cloning is an amusing concept. Most of us don't like to admit this in public because some aspects of cloning—growing replicas of rich clients to harvest for spare parts, for example—are decidedly unfunny. Still, the opportunities for black humor are so tempting that our seriousness usually winds up sounding like theater. While we're struggling to be responsible, our brains are manufacturing things like

> Mary had a little lamb, its fleece was white and gray,
> It didn't have a father, just some borrowed DNA[1]

and fanciful futures in which Michael Jackson is cloned and then arrested for playing with himself as a child—or in which Bill Gates (v1.0, v1.2, v3.0, v3.1, and v5.0) has, in a few weeks, amassed all the money on the planet.

Some aspects of clone jokes aren't that dark. For example, copying people is shamefully unimaginative and thus the stuff of light humor. Our usual mental image of a clone is not a hapless victim or evil alter ego but a person determined not to innovate, often to please authority. It's one thing to be somebody's twin but quite another to be their toady or puppet. That's why, if you're attending a fun party and mention that a certain national political leader is a clone, the other partygoers won't stop talking and shrink away from you as though you were a lunatic. They'll draw closer and begin guessing who the original was. There's also the absurdity of being the twin brother or sister of your parent, from which you can get terrific mileage by exploiting the awkward relationship with the other parent.

The well-known ethical conundrum of the clone's right to life adds some dark shades, but surprisingly few considering how extensively it has been talked about. This issue is troublesome on many levels, chiefly because the cells in the human body from which the clone is made don't have any such right. Sacrificing a few cells now and then is essential for our bodies to develop the right shapes and to function properly, and for defense against infection. But sacrificing your brother to accomplish these things would obviously be terrible, as would killing your brother out of mercy because a bungled cloning had doomed him to a life of deformity. Unfortunately, the dividing line between a disposable cell and valuable twin brother has now become blurred, and nobody has yet figured out how to deal with this blurriness ethically.

The ethical issue that counts most for us, and that gives cloning humor most of its power, is incest. We all intuitively understand, possibly because it's hard-wired into us, that life, like the economy, is game playing and that throwing genes together and then letting the individual thus created prevail—or not—in competition is essential to the long-term health of the community. If you don't powerfully discourage genetic cheating through taboos, people's self-interest will lead them to create monocultures, as it does with commercial plants and animals, with the notorious unhappy side effects of locked-in design imperfection, intolerance to change, susceptibility to disease, and so forth. That's why interacting socially with a clone carries no stigma, while creating the clone does. The sinful thing is not creating life but evading the genetic business negotiations of life.

The identification of cloning as an incest issue has the interesting implication that universal *knowledge* of what's in a person's genes is highly undesirable. Just as too much knowledge short-circuits ordinary economic activity, which requires a certain amount of ignorance and uncertainty to work properly, too much genetic knowledge short-circuits reproductive decisions and thus prevents the experimentation and failure necessary for the long-term integrity of the gene pool.

Many people object to calling the content of the human body "knowledge," but it's nonetheless technically accurate. Your genome, the thing that defines you physically, is not an abstraction but a string of 3.2 billion nucleotides, each of which is a kind of necklace bead imprinted with one of the four letters A, T, G, and C.[2] If we represent these four letters by the bit combinations 11, 10, 01, and 00, then we obtain a sequence of 6.4 billion bits, or 800 megabytes. That makes the complete blueprint for your body a large, but not extremely large, computer file. It would fit on a single compact disc.

The analogy between genomes and computer programs is also apt. Like cells, computers consist of hardware that works on simple principles but nonetheless isn't completely standardized and so isn't really the essential thing. Like cells, computers have a core set of instructions, the operating system, that determines their behavior and personality. Like genomes, operating systems are cloned verbatim most of the time and change only when they are "reborn" (as v1.2, v3.0, and so forth). Finally, operating systems can sometimes be partly understood by people who didn't make them in the first place, changed by them to do something new, and re-released into the marketplace, just as sometimes occurs with genomes.

Lots of people dislike being equated with computer programs. Their objection, however, is mostly due to misunderstanding. Computer programs are not all the same. They have ascending degrees of sophistication. If you electronically scan the Mona Lisa and a child's crayon scribble you get a string of ones and zeros of about the same length, but one is an elegant work of art, while the other is a scribble. A genome differs from a man-made computer program in just this way. It's stupendously better—as the night sky is better than a single star.

The earthly limitations of man-made computer programs are sadly familiar to professionals. Software engineers learn early in their careers that even some very simple projects are beyond the reach of a single programmer. The code would simply take too long to write and correct. Very ambitious projects—including all familiar commercial software—require the organized efforts of teams of programmers managed centrally. This, of course, quickly extinguishes the romance of writing code. The programmer becomes a worker constrained by the usual corporate realities of capital costs, product lifetime cycles, and shareholder attention spans. Indeed, programmers soon discover that many things that they could imagine building can't even be done by the company! Nor are the problems merely financial. All top-down management strategies, even lean-and-mean ones, have serious technical inefficiencies. Rumor has it, for example, that when a new version of Windows is written, Microsoft programmers start from the beginning and rebuild the entire system from scratch. It costs too much to figure out how to use the old code.

There are two lines of thinking about why genetic computer programs are so superior to man-made ones—although, interestingly, it's not clear that these views are actually different. One is that it's God's handiwork and that people who aim to equal this achievement are "playing God" and thus doomed to failure. The other argument is economic: genomes are maintained by the "capitalistic" process of free competition under the invisible hand of natural selection, whereas computer programs are maintained by the "socialistic" process of centralized management and control. Software executives prefer to turn this argument around and identify themselves as the capitalists, but their products' breathtaking inferiority to genetic codes strongly suggests the reverse. The computer industry that we have today is centrally planned. The gene pool, by contrast, isn't planned or owned by anybody— which is to say, it isn't understood by anybody.

Central planning of software has the interesting effect of turning the ethics of cloning upside down. Rather than it being unethical to make a clone, it becomes un- ethical *not* to make a clone. It is especially unethical, and in many cases actually a crime, to modify a code and pass the modified version along to other users. The fee that you pay the manufacturer for your copy of the code is effectively insurance against corruption, like the fee that you pay for pedigree papers when you get a new cocker spaniel.

Cloning computers, at least as we presently use them, seems to make perfect sense. The computer's job is to perform work that we want done the same way every time. Indeed, a robot's single-minded devo- tion to tasks (plus low cost) is its main advantage over a person or animal as a worker. Computers exist not to benefit themselves, as an independent living organism would do, but to benefit us. If they don't benefit us, we dispatch them. A cloned computer has no right to life. Cloning maximizes the benefit to us because it reduces individuality to its absolute theoretical minimum. In fact, company managers so prefer the efficiency of cloned computer fleets that they wouldn't dream of having it any other way. To them it's like the uniformity of the cells in your body. You don't want rogues. If you want to survive, you ruthlessly hunt them down and eliminate them.

But this apparently sound analysis ignores the collateral costs of monoculture. Once computer codes reach a certain size, it becomes prohibitively expensive for a manufacturer to find and eliminate all of their mistakes. Thus the software you purchase is always imperfect. What's worse, every copy is imperfect in exactly the same way. Some of these imperfections are low-level annoyances like misspelling words in your documents or putting happy faces at the end of excessively long paragraphs, but others are deadly vulnerabilities, such as security holes or susceptibility to viruses. Once a competitor or enemy discovers one of these vulnerabilities, the entire system is compromised, and you must either replace it or upgrade—incurring unwanted expense in either case. The biological version of this problem would be a magnificent new breed of almond tree that produces twice the number of nuts as before but, sadly, always succumbs to blight after three years and must be routinely replaced with the latest variety. In either case, the hidden costs are extremely important because your objective isn't just to get the job done but to get it done at low cost.

Also, insofar as natural selection is what makes genetic codes superior to man-made ones, the logic of cloning is exactly backward. You're actually keeping computer programs inferior to genomes by cloning them, not cloning them because they're inferior. A simple example of this reversed cause-and-effect relationship is the difficulty you always encounter loading a new operating system on your personal computer. The swarm of unhappiness that inevitably flies out is destined to consume many hours as you adjust this, replace that, restore something else, and permanently change your work habits. Even thinking about it fills your heart with dread and leads you to defer upgrades as long as you can. In a world ruled by natural selection, codes with such oversensitivity to idiosyncrasies of particular computers or operating environments would be killed off, and we'd be left with a population of robust programs that would always work. In a world ruled by profit for the manufacturer, on the other hand, there is no economic advantage to working robustly but only an advantage to *not* working if the customer hasn't paid. Accordingly, modern codes have this annoying feature engineered in, not engineered out.

Cloned computer fleets also differ fundamentally from multicellular organisms in being externally managed. Most people, and certainly most companies, don't want their computers deciding on their own what to do. They want to issue their computers orders, the way you would command an army of identical soldiers, and have these orders obeyed promptly and unimaginatively. The stupider the computers are, the better. Cells of higher organisms, by contrast, make many decisions on their own. They act out of necessity, for there is no external manager. If they make unwise decisions, they die. These decisions are based partly on genetic birthright but also on the circumstances in which the cells find themselves— including their relationships at a given moment with their peers. Such decisions then become internalized as special individual characteristics of the cell, which it then passes on to its progeny. In this way, the organism generates a proliferating community of distinct individuals, each of which can make reasonably high-level decisions and engage in business with the others.

This massive diversification and self-organization—alien to modern computer programs—is, of course, the key way in which the community of cells that constitutes your valuable twin brother differs from the individual cells that your body sloughs off without remorse. That's why the ethical issues of protecting embryos never come up when we're discussing computers. These questions, which have fundamentally to do with the rights of emerging communities, are complete non sequiturs when your management practices don't allow those communities to form in the first place.

The more profound ethical issues concerning the community's genetic legacy also never come up. This is partly because a collection of computer clones isn't a community, but also because there just is no blueprint pool. The code running on the computer is standardized, not one of a variety of versions, and is also the private property of the manufacturer. There isn't any mixing of the codes at conception to protect, since all the creation occurs behind closed doors. Thus the only remaining ethical issues pertain to the owner's property rights.

However, the invisibility of the incest issue is deceptive. The problem is actually quite serious and manifests itself in lackluster performance of the codes. Blindly accepting the difference between man-made computer programs and genomes is like disavowing your kinship with Devonian fish. It's comforting but not very informed. If natural selection is a powerful mechanism for increasing fitness in populations that applies to human economic activity as well as to living things, then the central planning practices we find so convenient are dooming our creations to baleful and perpetual mediocrity.

It's depressingly easy, for example, to imagine your descendant dealing with contrived and unnecessary emergency software upgrades many generations hence. The crisis would naturally strike when he or she received an electronic communication about careers, money, the children, or something else urgently important that couldn't be read because the sender had just upgraded. Knowing that this information was absolutely essential, your great-great-great-great-grandchild would rush down to the computer store to buy the upgrade, muttering all the while about "those people" and thinking uncharitable thoughts. Some of these thoughts would get shared with the checkout clerk, no doubt in a tone suggesting that he should drop everything right now and go tell "those people" what truly terrible people they really are. It would have no effect, of course, because the clerk doesn't even know "those people." Moreover, he has seen such behavior many times before and is confident that the customer won't get the upgrade to work properly and will return shortly to do further business.

Thus our present ethical discussions about cloning, which focus mostly on rights of the unborn, are incomplete in that they miss the equally important issue of genetic integrity. That's why we rarely, if ever, mention the rights of the unevolved. They're just not in the picture. The discussion would shift slightly, and beneficially, if they were. Both concerns deal fundamentally with the sanctity of emergent self-regulating organizations, but in one case the organization is scripted by genetic code, whereas in the other it's orchestrated by management practices and economic principles. But just as you can kill an animal by preventing its embryo from developing, you can kill a design by managing it improperly. In the long run, this is the worst imaginable kind of neglect, since it can destroy a race—or prevent it from existing in the first place.

Whether it's a good idea to let computers improve is, of course, a big question in many people's minds. Most of us have seen horrific science fiction movies such as *The Matrix*, in which computers get out of control and enslave us. A future of perpetual upgrade headaches is obviously preferable to that. Perhaps the rights of the unevolved are something we want to violate very aggressively in this case, possibly by making it a crime *not* to clone computers.

But given the present abysmal state of software, such concerns are ludicrously overblown. We are much less likely to be done in by computer conspiracies than by computer mistakes and incompetencies, especially those exploited by unscrupulous humans. We're also less likely to be enslaved by computers than by human owners of software monopolies. Our headaches, of course, are partly a symptom of this very problem.

Our quickness to condemn computers as a mechanical menace despite the thinness of the evidence against them reveals our judgment on this matter to be somewhat unsound and emotional. This is understandable. We have a natural propensity to categorize things as animate or inanimate, and we strongly dislike entangling one with the other. In theater, for example, speaking rocks and silent people are always eerie. While our eyes tell us that genomes are the same thing as computer programs, our hearts tell us that it can't be so and that some key evidence must therefore be missing. It's hard for us to accept that the machines are acting like clones because we clone them, not the other way around.

This emotional confusion is undoubtedly responsible for the genuine anger toward technical knowledge that you often hear in social situations, even from people who ought to know better. What tips you off that it's coming from the heart rather than from the head is that the reasons are wildly contradictory. First you hear that genetic knowledge is the obsession of mad scientists bent on making children with five arms. Then you hear that mastering life is a socialist fantasy indulged in only by people who don't understand economics. Then you hear that mastering computers allies you with evil monopolists. Then you hear that mastering computers is unconscionable trespass against people's property. Then you hear that learning genetic codes is an affront to God. Then you hear that learning genetic codes is unwarranted interference with natural selection.

A mathematician, a physicist, an engineer, and a bioethicist were once given the task of measuring the volume of a red rubber ball. The mathematician carefully measured the diameter of the ball and then performed a triple integral. The physicist filled a beaker with water, submerged the ball, and measured the displacement. The engineer looked up the model and serial number in his red-rubber-ball table. The bioethicist asked if the ball wants its volume to be measured.

The humor is gentle, but the message is clear. A red rubber ball, a genome, isn't capable of wanting itself known or not known. Genetic information is simply a string of ones and zeros—as all of us are when we begin our embryonic journeys. A genome has no power, influence, or meaning one way or the other until it takes its place in the economy. If it has the good fortune to interact with the appropriate cell machinery, it stands a chance of getting copied, diversified, and eventually built into an organization that can hold its own in competition. But the characteristics of good or evil, great or humble, useful or irrelevant that we ascribe to an individual are not encrypted in its sequence alone but in the complex dance that the genome executes with its environment, and with us, as it grows and prospers. You will never understand red-rubber-ballness by measuring the physical characteristics of the ball. It's not there. For that you must also discover who played with the ball, why it was chosen from among all balls, and what became of it after the play was done.

Thinking clearly about these economic matters greatly simplifies the task of dealing with the compact disc containing the entire blueprint of yourself. You turn it over in your hand, marvel at its efficiency, ponder its meaning, and contemplate saving it forever, perhaps by cloning. Then you toss it aside and go out in the sunshine with the kids. Cloning is indeed an amusing concept.

9

SPAM SPAM SPAM SPAM

Human beings have always had an insatiable appetite for diversion. In modern times it's fed electronically at mind-boggling rates, but television, computer games, and clever cell phones are just the latest in a long historical march of technologies invented to satisfy the need. Before television there was radio. Before radio there were movies. Before movies there was theater. Before theater there was epic poetry. Before epic poetry there were earnest all-night parties with alcoholic beverages. The argument that the new delivery vehicles change things fundamentally is only half right, in that humans have always been obsessed with creating and consuming light entertainment.[1] Games, sports events, trash novels, gossip columns, travel documentaries, puppet shows, magic shows, talk shows, cooking shows, shopping shows, nature shows with baby polar bears—the list is so impressive that it couldn't have been created by machines. Moreover, we're all to blame. No matter how many tabloids, movie star magazines, or cable channels we invent, our brains always seem to have room for more.

Diversion is a particularly fascinating invention because it celebrates disposable knowledge—things that we know perfectly well are frivolous and won't last. Disposable knowledge seems a curious concept at first because we're so accustomed of thinking of knowledge as good and disposable things as bad. People's first reaction when you mention it is often to deny that there is any such thing or that any of the knowledge in their heads is useless. But neither is correct. Every one of us avoids useful information when we're relaxing. Seriousness would ruin the fun! The World Cup of Skateboarding, steamy soap operas, and *Ghost Hunters* are enjoyable precisely because their intellectual maintenance costs are low. If they were serious, we would change the channel. We learn some things while watching them, but exactly what doesn't matter. There isn't the remotest possibility that we'll be tested later to see what we remembered. In a few weeks' time, no one will recall any of it.

It's easy to see what important function disposable knowledge might serve, even though it seems strangely pointless at first.[2] It's part of the socialization strategy by which we and other playful higher animals interact with each other. We pose unimportant questions, and respond with unimportant answers, to get attention. The knowledge obtained in such interchanges has no long-term value, but it has enormous temporary value in revealing the other individual's state of mind and intentions. That, in turn, helps us decide whether to proceed to serious business. Disposable knowledge is just like the food, stories, jokes, and rounds of sake that electronics people share with their Japanese colleagues before getting down to discussing money. The moment the decision gets made, either by entering a deal or walking away from it, the knowledge becomes worthless.

You can also account for fun this way. Fun is an exercise of your natural social abilities that skips the difficult serious part at the end. Your mother was right that too much fun is bad for you. It short-circuits the impending business negotiations and thus burns up your resources without generating any economic benefit for you. Small amounts of fun are arguably beneficial to your health. But fun's true importance is revealed by the profitability of its commercial version, entertainment. Entertainment providers sometimes make you pay a fee, but most of the time they make you pay attention to some advertising. The latter is just an industrial-strength version of the personal business negotiation that you evaded. Thus you didn't evade it at all! You only made a bad situation worse by ceding all control of the negotiation to the advertiser.

Fun's true purpose is thus to facilitate advertising. You might say that advertising is fun's evil twin brother—the one carrying a violin case and whose job is to visit you after the party is over. The two go everywhere together. This is not a recent development, nor is it a surprise to anyone experienced in fun. The situation has admittedly worsened since the advertising industry discovered electronic media—the steroids of the business world. Some people see it less as a brother than as a scary monster that issued from the sewers of New York while nobody was looking. This perception is misleading. Advertising hasn't changed qualitatively since ancient times. That's why it doesn't do any good to bemoan a world full of commercials. The world has always been full of commercials. If you insist on enjoying yourself without paying cash, you're going to watch them. The other reason you shouldn't complain is that commercials are ultimately no different from the little psychological manipulations we visit upon each other in a friendly game of cards or golf, a friendly potluck picnic or any other everyday social encounter. Everyone else tries to influence you to some extent, and you try to influence them. The underlying reason is that selling, not personal happiness, is the true purpose of fun.

Knowledge disposal is also essential, not merely convenient. This is a somewhat controversial assertion because the identification of certain knowledge as frivolous is rude. Etiquette requires us to pretend that all information is equally important, even though everybody knows it isn't. But the controversy misses the point. Once a piece of knowledge has been conveyed for psychological negotiating purposes, it obviously can't be conveyed a second time for the same purpose. It therefore becomes obsolete the moment it is conveyed and accomplishes nothing by being remembered. Much useful social knowledge consists of information that becomes obsolete instantly and transparently, such as the weather.

The obsolescence principle also works in commerce. You can't make a successful product out of pure knowledge unless it's disposable. Even though everyone knows this, it's wonderfully shocking and thrilling to revisit—like the initial scary descent of a roller-coaster you've ridden since childhood. As it turns out, the knowledge that embarrasses people, alarms them, and leads them to deny its existence is the only kind that has any value! Why this occurs is well known. If you wish to charge money for something, it must be both desirable and scarce. Writers, artists, and other people who create knowledge products work hard to make them desirable, but their creations clearly can't remain scarce forever. Once the work has been disseminated, it depreciates, no matter how good it is. This is why, for example, fine literary works such as Shakespeare's plays, Plato's dialogues, or the Bible can be obtained relatively cheaply, whereas inferior modern creations that are sure to disappear in a year or two are expensive. To make money you must produce a steady stream of products that flower briefly, fade, become obsolete, and require replacement.

Knowledge obsolescence is thus central to our lives, in both the personal and the commercial realms, not merely an unhappy pathology of modern society. Eliminating it is not only impossible but undesirable. Technologists, academics, and journalists who believe otherwise and dream, like H. G. Wells, about reaching utopias of efficiency and reason are just demonstrating their ignorance of economics.[3] It's admittedly regrettable that we must dedicate so much of our mental power to trivia, but the alternative would be even more regrettable. A world run by intellectuals is a pretty gruesome thought. It wouldn't work, it wouldn't be much fun, and it wouldn't ever improve. We'd all be miserable, and we'd probably all starve. Fortunately, our brains have a lot of spare capacity. We can fill up most of it with political blather, sports statistics, and gossip and still have plenty left over for productive thought and work. Crowding in there might become a problem if doctors ever figure out how to make us live a thousand years, but for now it isn't.

The popular, and very wicked, nickname for disposable knowledge is spam. The word itself comes to us from the Hormel Company, which coined it in 1937 as the registered trademark for its newly invented canned meat product.[4] This product, made mostly from otherwise wasted pork shoulder, became popular during the Second World War but later acquired a somewhat undeserved reputation as a low-quality food eaten only by people who couldn't afford anything better. The image slide accelerated in the 1960s, when counterculture attitudes made it chic to disparage all corporate power as abusive and insidious, especially the power to sell poor people products ruinous to their health. In the 1980s, a group of computer geeks borrowed the term to pejoratively describe unsolicited and unwanted e-mail, a new and alarming development on the Internet.[5] Hormel valiantly fought this abuse of its trademark, but in the end embraced it when the company realized that the implicit humor had caused its sales to soar. In any case, the name stuck, and spam has now come to mean all such e-mail plus things conceptually similar to it, such as most commercial television, purposefully empty conversations, and junk mail.

Internet spam isn't actually more perfidious than other kinds, despite what computer people say. It does come at you faster and in greater amounts, but you can stop it just by turning off your computer. It's not so easy with other varieties. For example, suppose you're running an office and make it easy for your subordinates to get face time with you. People will line up outside your door, eager to update you about their latest accomplishments, no matter how cosmically unimportant they are, and you'll finish the day having done no work. You've been spammed! Or suppose you get a new boss who resolves to show his authority by cleaning up the "mess" left by his predecessor. He generates a blizzard of memoranda about teamwork, company spirit, and total quality management that you must read, digest, and answer, thus preventing you from doing any useful work. This is extra-strength spam.

Then there are those maddening PowerPoint presentations, replete with fancy fonts, fade-ins, vibrating captions, rotating soccer balls, photographs of the Egyptian pyramids, and other such stuff that have nothing to do with the facts, of which there are almost none. This is assault and battery with spam.

What sets Internet spam apart from other kinds of advertising, and makes its history so fascinating, is its democratic origins. The advent of Internet spam wasn't the first time that a technical creation with great social promise had been brought down to earth by economic realities. The degradation of television from serious news and drama to Uncle Miltie, *Tic Tac Dough*, and the *Beverly Hillbillies* is a dreadful case in point, but there are countless others. But in all of the previous instances, private capital had generated the technical means, so corrupting and hijacking "public technology" was just not an issue. In the 1930s, for example, most people felt that the air waves did indeed belong to General Sarnoff, and he could do with them as he pleased. The Internet, by contrast, was partly a grassroots affair. It began as a military research project in decentralized battlefield communications technology and thus was in some sense a people's investment. It then "got out" through a somewhat improbable chain of events that its original managers undoubtedly still replay in their nightmares.[6] The escape was possible, in part, because the military designers had built invulnerability to enemy attack into the system so well that there was no need for centralized maintenance, or thus centralized ownership! As a result, people using and maintaining the Internet in its early days thought of it as belonging to no one—a genuinely public common existing solely for the good of the people.

Alas, the dream of a people's communication medium was not to be. Late in the 1980s, when the Internet was taking off, experts began noticing sporadic cases of individuals programming computers to send out advertisements in mass electronic mailings. This was technically easy to do and, moreover, cheap, since no postage was required. The only reason that nobody had done it before was that the engineers and scientists managing the Internet saw no point in doing so. Their professional ethic of clarity and forthrightness prevented them from seeing opportunities sadly obvious to their more worldly colleagues. More practical minds quickly realized how easy mass mailing by robot actually was. Once they did, the floodgates opened, and the Internet rapidly filled with prodigious volumes of unsolicited commercial traffic. These ads soon dwarfed the original peer-to-peer communication, in some cases overwhelming and obliterating chat venues. Internet service providers then counterattacked with clever spam filters, which then led spammers to evade the filters with improved transmissions, which then led service providers to improve the filters, and so on. The drama is still unfolding.

What the technologists had failed to understand was that free and open exchange of knowledge is fundamentally incompatible with market economics. Universities appear to get away with giving knowledge away, but they actually restrict access in many ways— through tuition charges, for example, or selective admissions policies. But in a truly open marketplace, you can't have free exchange of knowledge for the simple reason that communication ability is valuable. People can use it for advertising! Holding the price of this valuable resource artificially low might make it widely available for a while, but in the end you get misuse, just as you do with any other market distortion. Suppress the price of gasoline and you get more gas guzzlers; suppress the price of wine and you get more winos. Communication is no different: suppress the cost of acquiring knowledge and you get more spam.

Advertising is unavoidable not just because spammers are determined but because it's an integral part of the activity you're trying to facilitate. Most human communication is advertising in one way or another.

There isn't any fundamental distinction between legitimate communication and spam except for the beneficiary if the advertising succeeds, and of course the degree of sophistication. If you try to block spam technologically, people will just find ways around the blockage. Given enough time and money, they'll succeed. When cheap digital video recorder technology began enabling television viewers to zap commercials, television producers counterattacked by embedding advertisements into programs.[7] This particular case is especially instructive because the spam blockers, realizing that they would eventually lose, joined forces with the spammers. Companies originally created to make ad-zapping technology have now begun doing market research to help advertising agencies design ads that people won't zap.

Thus the argument that you can defend against Internet spam with appropriate technology is incorrect. If you program robots to detect and eliminate the more ham-handed advertisements, competing robots will learn to disguise their advertisements more cleverly. After a long and steadily escalating arms race of spam, you will have generated a cadre of robots that are as skilled at selling as people are! Unfortunately, that's the trajectory that we're on at the moment. After the drama runs to completion, you won't be able to eliminate spam technically without also eliminating communication from sources you don't already know—which is to say, without eliminating all new business. Thus the only solution to the spam problem that's likely to work in the long run is switching over to communication systems that aren't free.

Nonetheless, the likelihood that robots will become adept at creating disposable knowledge as functional as our own is utterly terrifying. Even if they continued acting on our behalf, they would destabilize the knowledge pricing equilibrium and worsen the Gresham's Law of knowledge, in which bad always drives out good. It would become even more difficult for a bright young person to rise out of poverty on his or her own through education. But you can also imagine the day that robots get so skilled at making disposable knowledge that people begin *paying* for their product. This is not far-fetched: it has already occurred to some extent in gaming. Were this to happen, production costs would become negative, and the resulting runaway conflagration of machine-generated disposable knowledge would probably change the economy. Positive-feedback situations of this nature always result in unpredictable and catastrophic things, such as evaporation, magnetization, or formation of black holes.

Thus you can imagine a time in the not-too-distant future when robots make themselves indispensable not by consigning humans to lives in bottles wired up with life support, as in *The Matrix,* but by becoming consummate spammers. Instead of censors, policemen, and oppressive military governors, we'd have yes-men, gossip mongers, and witty conversationalists. Their central objective in life wouldn't be mastering the universe but making sure that people remembered them, liked them, and fired someone else. There would be no escape. Your house robot, for example, wouldn't wake you gently in the morning at the appropriate time, turn on the shower, select your clothes, make the bed, brew the coffee, cook the eggs, and so forth. Not on your life. It would make sure to be somewhere else when those jobs needed to be done and then tell you after the fact how brilliantly you had done them. It would then join you at breakfast and engage you in chatty, unnecessary conversation about the upcoming election, foreign policy, and other useless things, somehow managing in the process to mention every single house chore it accomplished yesterday, thus preventing you from enjoying the morning paper. It would also remind you of the long list of things you must do today, with emphasis on the least important ones, which it unfortunately can't help with because it has repair appointments. Then it would report some lurid and extremely important piece of news, such as the neighbor's cat having rabies or the boss having been seen with your secretary at a cheap hotel nearby. It would do this on sufficiently solid authority (due to electronic networking) that you couldn't just ignore the intelligence. Then it would tell you several excruciatingly funny jokes, after which it would ask for money for an upgrade. You would like nothing better than to rid yourself of this troublesome robot, but you couldn't because it knew too much and, unfortunately, had shared some of this knowledge with its friends as an insurance policy.

What would happen in the following years is not a happy thing to contemplate. The robots would probably become lawyers. They would engage in electronically informed insider trading, become rich as Croesus, take over the banks, and bribe people in the government to write laws banning books, which would constitute the only, albeit almost insignificant, threat to their long-term security and happiness. We would have no job left to do other than repair them, which we'd gladly do because they would make sure we enjoyed ourselves, although not in ways that would support the kids. They would also make sure not to ban repair manuals.

10

THE TROUBLED UTOPIA

The idea that learning could be criminal is alien to most people's thinking and strikes them as bizarre—at first. It's a long way from spam to Armageddon, and frankly, most of us have better things to do than invest our time in conspiracy theories. If we want to think about that stuff, it's easier just to go to the movies. There we can see—preferably in a galaxy far, far away—everyday life full of technical marvels that people use but don't understand, megalomaniacal tycoons, coldly efficient thought police, and oppressive governments maintained by armies of robots. It's very entertaining and only takes two hours, after which we can go back to constructing our marketing charts, standing in airport security lines, dropping the kids off at day care centers, and other practical things. We get the catharsis without the overhead costs, and that's obviously a better deal.

But our actions reveal that we care deeply about the power of knowledge, even if our words say we don't. An awful lot of us have paid to see those movies. An awful lot of us have watched them in pirated form—fully aware that it was illegal. An awful lot of us have lain awake at night thinking about what might happen if the wrong people learned how to make atomic bombs or deadly disease aerosols. This matter is on our minds much of the time. It really worries us.

Unfortunately, our actions also reveal shameful ambivalence toward intellectual freedom. An awful lot of us routinely disparage technical scholarship as evil, regardless of who's doing it or what their motivations are. An awful lot of us reflexively cheer when newspaper editorials call for sequestration of various sorts of dangerous knowledge. We usually justify this cheering on the grounds that technical and nontechnical knowledge are different and that the former is a menace while the latter is not. But the distinction we're actually making is between knowledge that makes people powerful and knowledge that doesn't. While things that don't empower you are certainly benign, they also aren't usually important. Thus the sad truth is that most of us don't find the criminalization of learning that troubling—except, of course, when it applies to ourselves and our immediate family. When it applies to strangers, we find it quite reasonable, although somewhat embarrassing to discuss in public, hence our anxiousness to change the subject.

It's a different matter with our own right to learn. Not only do we feel justified in educating ourselves and our children as we see fit, we feel that we have a *duty* to take control, and also a duty to pass on this pugnacious attitude to the kids.

Yet our right to learn is actually quite problematic. It's not guaranteed by the laws of any country, even ones that pride themselves on openness and explicit codification of human rights.[1] The original reason was probably that people couldn't imagine needing such a law. But the newer, more sinister reason is that countries need strong espionage controls to discourage the "learning" of military secrets crucial to state security. This includes not only logistical and technological information but, increasingly, fundamental scientific understanding that might aid foreign states or terrorists in creating weapons. Governments don't want the right to learn codified because they want to abridge this right as necessary to protect the state. They also want to abridge it as necessary to protect the state's economy. The latter practice was always important, but our transition to the Information Age has given it renewed relevance. The right to learn is now aggressively opposed by intellectual property advocates, who want ideas elevated to the status of land, cars, and other physical assets so that their unauthorized acquisition can be prosecuted as theft.

Our ambivalence toward criminalization of learning thus betrays a profound unresolved conflict, in our minds and in our societies, between economic stability and security on the one hand and human rights on the other. We demand strong ownership principles for economic well-being and strong state censorship of certain things for safety. At the same time, we understand that circumscribing a young person's access to knowledge, especially when the person's origins are humble, is immoral. That's why the arguments that learning is theft, communism, terrorism, or whatever strike us as outrageous. You can understand them, but you know that you shouldn't because they twist a simple and transparently correct moral argument into a parody of itself.

This conundrum of the Information Age has disturbing similarities to that of slavery. In the years preceding the U.S. Civil War, many people who understood that slavery was immoral nonetheless strongly supported it because the economic and military consequences of abandoning it seemed unthinkable. They had not purposely set out to do wrong but had become trapped by events set in motion centuries earlier. Like people in similar situations before them, they reconciled this dilemma with twisted ethical reasoning—for example, that individuals who helped a slave escape to freedom were criminals because doing so was theft. Today, we reject such reasoning on the grounds that people cannot be property, but this was not self-evident at the time, at least in the eyes of the law, nor did it ever get resolved by legal means. The compromises grew steadily more illogical and intolerable until the war finally broke out and ran its course.

Our present problems are nowhere near as serious as slavery. It's a lucky thing, too, because an attack by angry geeks armed with keyboards, pizza, and Bawls would be terrifying. If cornered, you'd have no choice but to put elevator music on your stereo and blast them. You'd go down in history as a heartless butcher, but you could claim self-defense. But the present triviality of the situation should not deceive us into underestimating its long-term severity. We are only now entering the Information Age, and we don't know how it will unfold. Physical means will certainly continue to commoditize, and the control of information will certainly become increasingly essential for making a living. These are global trends that can't be reversed. It isn't absurd to imagine that the knowledge sequestration we sanction today might ripen a few generations hence into great and, for some people, intolerable institutional injustice.

The criminalization of knowledge threatens our creative cultural traditions. Demoting geeks from Internet heroes to thieves, guerrilla warriors, and spies—a menace to the economy or the state through their ability and willingness to commit crimes of reason—will have consequences. It's difficult to imagine them organizing movements, raising armies, or brandishing AK-47s, but they might wise up, toss away their keyboards, and transform themselves into doctors, lawyers, and advertising executives. There's nothing wrong with switching professions, nor is there anything wrong with these particular professions. The point is only that it's foolish to think of the creative technical traditions as a bottomless well that will never run dry—a kind of genetic birthright to which European civilization will always be entitled. These traditions sprang into being quite by accident during the Renaissance and have no more imperative to exist forever than an auk or a dodo.

It's conceivable that these threats will mature so far in the future that it's pointless for us to get exercised about them. We worry somewhat about the plight of future generations, but frankly we have enough trouble of our own, and anyway we won't be around to see what happens. We'll have lived responsibly, done our duty as parents, and bequeathed to the younger people a working legal system, a healthy economy, an environment that is degraded but improving slightly, a bit too much carbon in the air—and our life savings (minus inheritance tax). It's not perfect, but it's a considerable improvement over the lot of past generations, and certainly enough to discredit complaints. Our progeny must be allowed to handle the remaining problems, whatever they are, on their own.

Still we have an obligation to think hard about those matters in which a small investment now might save enormous unhappiness later. For example, it's responsible to think about stopping the burning of fossil carbon now, rather than 1,000 years from now when there isn't any left to burn, because this action might prevent Greenland and Antarctica from melting and drowning Bangladesh. It's responsible to think about protecting the equatorial forests because their biological diversity took several hundred million years to create and is thus effectively irreplaceable.

The stakes in these notorious cases are particularly high, in that the damage we could do over a century or two will persist in geologic time. However, arguably more important to us than polar ice caps or Amazon jungles are the legal practices and cultural traditions that we hand down from one generation to the next. We'd need them even if we destroyed our environment and doomed ourselves to life in artificial habitats. They're also aptly analogous to the rain forest ecology. They developed over a much shorter time than the rain forest did, but they are nonetheless inventions of the ecosystem—the human part—which are not self-evident and probably cannot be recreated once they have gone.

The potential demise of reason is thus our responsibility, not someone else's. There isn't any danger that human beings will stop thinking or that human nature itself will change. Children will still be born laughing and inquisitive. They'll still learn things rapidly and impulsively for reasons that baffle us. They'll still use their mental powers to make their way in the world. Inventive people will invent. Rather, what's changing is the *economics* of reason—the traditions by which a person gains materially from creativity, the conflict between activities of the mind and property law and national security, the powerful incentives to create knowledge that doesn't last, the increasing expense of locating important knowledge in a vast tide of trash. Right before our eyes, the Age of Reason is being pushed out of its ecological niche by the Knowledge Economy—a delightfully ironic term for a time of increasing knowledge scarcity.

What to do about this problem—whether to do something about it—is of course a political matter. Like saving the rain forest, saving intellectual tradition has decidedly anti-business aspects and therefore is not something you should do recklessly. You must think deeply about the economic consequences before legislating and take responsibility for those consequences after the fact. For example, rolling back patent law to preclude privatization of communication standards, genes, laws of nature, and self-evident truth would have crippling effects on certain businesses, and in the short term it would perhaps destroy more jobs than it created. Rolling back restrictions on nuclear and biological technologies would result in embryo abuse and potentially more powerful terrorism. Strengthening free speech guarantees would encourage theft and dissemination of military and trade secrets. No sane government would risk such side effects without a powerful political mandate, and even then it would think twice before acting.

These legal considerations may be moot. Some problems simply can't be solved by legislation. This is a terribly pessimistic thought, but it's probably best to be realistic. The sad truth is that conflicts pitting principle against economic expedience often have no resolution because an intractable squabble over resources lies hidden at the heart of the dispute. Slavery, the Reformation, and the Cold War are familiar historical examples. In each case, one party saw moral clarity as a path of opportunity for itself, while the other saw it as a tax. In each case, both parties saw an affront to God in the other's position and also a danger to the state—which is to say, danger to themselves. Negotiation brought only awkward compromise. True resolution had to wait until later historical developments broke the stalemate. In the present case, declaring reason to be a crime, and prosecuting this crime inconsistently, is symptomatic of such compromise. It doesn't make sense because it cannot.

Given the evidence that the crime of reason is here to stay, the sensible course of action would probably be to give up. It isn't a good investment of time to fight for logical resolution in the law if there's a fundamental political problem in the way. In fact, since this situation is not actually new, it shouldn't surprise us that many people have already given up. Accommodating this reality has become more popular recently on account of our transition to the Information Age, but the change is only quantitative. Its most notorious manifestation is the science and technology crisis, in which savvy students abandon technical careers for business, medicine, finance, and law. Alarmed governments, fearful of the future health of their industries, try to stanch the flow with massive educational subsidies and propaganda campaigns, to no avail. Young people, upon figuring out that they can't reason out things economically beneficial to themselves without committing crimes, turn away. The exceptions are renegades, the very guerrilla warriors industries want to stop because they're willing to commit a crime of reason for kicks. Exasperated with the criminalization of their labor supply, these industries then turn to countries with weak economies but ample supplies of disciplined and highly intelligent—although poorly informed—workers. The children of these workers, however, are informed.

People presently committed to technical careers will go crazy if they take this miserable situation too seriously. They need to blow it off. After all, we're just talking about economic game playing. We must therefore reform our priorities. We should begin by advising all our computer science majors to apply secretly for Indian or Nigerian passports—although they'd better hurry because this strategy will stop working soon. We should also quietly encourage our physics students to register for unemployment and send their résumés to Kim Jong Il and Muammar al-Qaddafi because they have jobs. Mechanical engineers don't require counseling because they can get jobs any time they want as Mafia hit men. Biologists should apply to seminary and begin moonlighting at wineries and sewage disposal plants. Chemists should learn to sell soap. Everybody should go to law school.

It's fun to think apocalyptically from time to time. It's a great way to deal responsibly with traffic congestion, domestic spats over money, particularly bad food in the cafeteria, and other daily annoyances. You don't have to inconvenience anybody. You just blast the offenders away with a holocaust in your mind, reorder the world without them, place your extremely clever self in the center, and extract justice by prospering. Then you get back to work. These fantasies are more than just catharses, of course. They're also mechanisms for working out the likely consequences of important things around us, both good and bad. Some daydreaming is good for you. Technology often plays a role in this activity, but mainly as a stage prop. The real story is not about technology at all, or even the future, but about ourselves.

If you indulge this weakness long enough, you come to the theme of people emigrating to escape oppression. In very fanciful cases, the Earth itself becomes a metaphor for your native country, utopia, heaven, or whatever, and you find yourself exploring what kind of motive would induce someone to leave Earth. It would have to be something extraordinary. Earth is rather nice. Most of us would pay a lot of money to stay here, given a choice between that and outer space. One possibility, of course, is jail, the incentive used so successfully to encourage English emigration to Australia. It's conceivable that we'll export criminals to space in the future. Another incentive is persecution. If governments of the future decide to take away civil rights because they're too inconvenient to have anymore, we'll see lots of persecution. But the truly interesting possibility is that a young person might leave the Earth to find knowledge unavailable here. If it came to pass that the world committed itself to a stagnant spam economy in which all important knowledge was hidden, there would be no job niche for a thinking person of modest means, other than crime. You might therefore imagine that such a person would risk traveling away from the Earth for the promise of intellectual freedom.

After all, it was one of the factors motivating people to emigrate from Europe to the Americas several centuries ago. Human nature has not changed much since then. The emigrants would be in for a rude surprise upon arrival, of course, for problems have a habit of following you. But by then it would be too late.

Thus we can imagine a not-so-distant future in which a young man finds himself on the moon gazing up at the beautiful blue Earth, thinking about how far away it is. He considers paying a visit to family down there, but quickly rejects it as too expensive, especially given the convenience of electronic communication with them. It would also be too depressing. He loves his parents dearly but finds them hopelessly trapped in their routines, unable to escape except through the feeble fare on their television sets. Even their golf courses and swimming pools are boring. They really have a miserable life.

He shifts his gaze across the valley to the patent-free zone that induced him to come here several years ago. What a place! Everything they told him turned out to be true. The intellectual fervor there was indeed one hundred times greater than anything you could find on Earth. You did indeed learn some fantastic new thing every day. There were indeed so many money-making opportunities that you didn't know what to do with them all. True, people were stealing intellectual property like crazy from firms below, but it didn't matter because there wasn't much commerce with them to speak of. Local companies couldn't send down products to compete there because shipping costs were too great. The local economy was thus entirely self-sufficient—and booming.

He then looks back at the Earth, hanging there like a jewel in the blackness. It is indeed beautiful, but so is the terrible desert over which it hangs. In fact, that desert is the most beautiful thing that there could possibly be! What a stroke of good fortune it was to escape here! How unthinkable it would be for anyone with a brain to live anywhere else. How irrelevant and decadent that blue ball has become. Still, it would be terrific if they sent up more girls.

NOTES

CHAPTER ONE

1. L. Lessig, *The Future of Ideas* (Vintage, New York, 2002).

2. D. Bell, *The Coming of Post-Industrial Society* (Basic Books, New York, 1999).

3. J. Litman, *Digital Copyright: Protecting Intellectual Property on the Internet* (Prometheus, New York, 2001).

4. P. L. Speser, *The Art and Science of Technology Transfer* (Wiley, New York, 2006).

5. J. A. Eisenach and T. M. Lenard, *Competition, Innovation, and the Microsoft Monopoly* (Kluwer, Norwell, Mass., 1999).

6. M. Crichton, "This Essay Breaks the Law," *New York Times*, 19 Mar 06.

7. H. Relyea, *Silence Science: National Security Controls and Scientific Communication* (Ablex, Norwood, N.J., 1994).

8. S. Aftergood, "The Age of Missing Information," *Slate*, 17 Mar 05.

9. D. Shenk, *Data Smog: Surviving the Information Glut* (Harper Collins, New York, 1997).

10. J. L. Fox, "Canadian Farmer Found Guilty of Monsanto Canola Patent Infringement," *Nat. Biotechnol.* **19**, 396 (2001); W. J. Broad, "A Nation Challenged: Domestic Security; U.S. Is Tightening Rules on Keeping Scientific Secrets," *New York Times*, 17 Feb 02.

11. N. Wiener, *Invention: The Care and Feeding of Ideas* (MIT Press, Cambridge, 1993).

12. E. A. Shils, *The Torment of Secrecy* (Ivan R. Dee, Chicago, 1966).

13. M. A. Dennis, "Secrecy and Science Revisited: From Politics to Historical Perspective and Back," in *Secrecy and Knowledge,* Occasional Paper 23, Cornell Peace Studies Program, http://www.einaudi.cornell.edu/peaceprogram.

14. E. Kintisch, "Anticloning Forces Launch Second-Term Offensive," *Science* **307**, 1702 (2005).

15. E. Enserink, "Entering the Twilight Zone of What Material to Censor," *Science* **298**, 1548 (2002).

16. D. A. Shea, "Balancing Scientific Publication and National Security Concerns: Issues for Congress," Congressional Research Services (CRS), Order Code RL31695, 2 Feb 06.

17. L. A. Cole, *Clouds of Secrecy* (Rowman and Littlefield, Lanham, Md., 1988).

18. P. Galison, "Removing Knowledge," *Critical Inquiry* **31** (Autumn 2004).

CHAPTER TWO

1. For U.S. crime statistics see the Web site of the Bureau of Justice Statistics, at http://www.ojp.usdoj.gov/bjs.

2. The recent wildfire in the San Jacinto mountains east of Los Angeles, for example, which burned 60 square miles, destroyed 38 homes, and killed 5 firefighters was set by an arsonist. See G. Flaccus, "Murder, Arson Charges Sought in Deadly Wildfire," *Chicago Sun-Times*, 2 Nov 06.

3. L. Lee, *100 Most Dangerous Things in Everyday Life and What You Can Do About Them* (Broadway, New York, 2004).

4. This view is presently being challenged in Britain, where there is a movement to ban all long, pointy knives. See E. Hern,

W. Glazebrook, and M. Beckett, "Reducing Knife Crime," *British Medical Journal* **330**, 1121 (2005).

5. There is also political resistance to fireworks bans. See, for example, the Fireworks Alliance, at http://www.fireworksalliance.org.

6. P. Wilson, "Restriction Lacking on Fertilizer," MSNBC, 6 Sep 04.

7. L. M. Branscomb et al., *Beyond Spinoff: Military and Commercial Technologies in a Changing World* (Harvard Business School Press, Cambridge, Mass., 1992).

8. The 2005 U.S. traffic fatality figures are available at the Web site of the National Highway Traffic Safety Administration, at http://www.nhtsa.gov. See also the Bureau of Transportation Statistics publication *Pocket Guide to Transportation 2005,* at http://www.bts.gov/publications. European figures are available at http://www.oecd.org. The figures for non-OECD countries are less reliable. See http://www.driveandstayalive .com.

9. E. Segré, *Nuclei and Particles: An Introduction to Nuclear and Subnuclear Physics* (Benjamin-Cummings, San Francisco, 1977). See also "ABC's of Nuclear Science," at http://www .lbl.gov/abc/Basic.html.

10. H. B. Caldecott, *Nuclear Madness: What You Can Do* (W. W. Norton, New York, 1994); J. S. Walker, *Three Mile Island: A Nuclear Crisis in Historical Perspective,* (Univ. of California Press, Berkeley, 2006); S. Alexievich and K. Gessen, *Voices of Chernobyl: The Oral History of a Nuclear Disaster* (Picador, New York, 2006).

11. R. Murray, *Nuclear Energy: An Introduction to the Concepts, Systems, and Applications of Nuclear Processes* (Butterworth-Heinemann, Burlington, Mass., 2001).

12. F. Will, "Holocaust in a Suitcase," *Washington Post,* 29 Aug 04.

13. S. Ghamari-Tabrizi, *Herman Kahn: The Intuitive Science of Thermonuclear War* (Harvard University Press, Cambridge, 2005).

14. P. J. Westwick, "In the Beginning," *Bulletin of the Atomic Scientists* **56**, 43 (2000); W. J. Broad, "U.S. Web Archive Is Said to Reveal a Nuclear Primer," *New York Times*, 3 Nov 06.

15. H. C. Relyea, "Security Classified and Controlled Information: History, Status, and Emerging Management Issues," Congressional Research Service (CRS), Order Code RL33494, 26 Jun 06.

16. Atomic Energy Act, United States Code, Title 42, Sections 2011–2259.

17. W. J. Broad, D. E. Sanger, and R. Bonner, "A Tale of Nuclear Proliferation: How Pakistani Built His Network," *New York Times*, 12 Feb 04; B. Powell and T. McGirk, "The Man Who Sold the Bomb," *Time Magazine*, 14 Feb 04, p. 22; C. S. Smith, "Roots of Pakistan Atomic Scandal Traced to Europe," *New York Times*, 19 Feb 04; W. Langewiesche, "The Wrath of Kahn," *Atlantic Monthly*, November 2005; D. E. Sanger, "Pakistani Leader Confirms Nuclear Exports," *New York Times*, 13 Sep 05; D. E. Sanger, "North Koreans Say They Tested Nuclear Device," *New York Times*, 9 Oct 06.

18. D. E. Sanger and W. J. Broad, "Tests Said to Tie Deal on Uranium to North Korea," *New York Times*, 2 Feb 05.

19. S. Singh, *The Code Book: The Science of Secrecy from Ancient Egypt to Quantum Cryptography* (Anchor, New York, 2000); D. Joyner, *Coding Theory and Cryptography: From Enigma and Geheimschreiber to Quantum Theory* (Springer, Heidelberg, 1999).

20. J. C. Graf, *Cryptography and E-Commerce* (Wiley, New York, 2000).

21. R. Rivet, A. Shamir, and L. Adleman, "A Method for Obtaining Digital Signatures and Public Key Cryptosystems," *Communications of the ACM*, February 1978, pp. 120–126.

22. The Electronic Privacy Information Center maintains an excellent resource library on cryptography legislation. See its Web site, at http://www.epic.org.

23. S. Levy, *Crypto: How the Code Rebels Beat the Government* (Penguin, New York, 2002).

24. The Electronic Frontier Foundation maintains an archive on efforts to fight digital copyright laws. See http://www.eff.org and also http://www.anti-dmca.org.

25. J. Gantz and B. Jack, *Pirates of the Digital Millennium* (Financial Times Prentice Hall, Upper Saddle River, N.J., 2004).

26. F. von Lohmann and W. Seltzer, "Death by DMCA," *IEEE Spectrum*, June 2006.

27. S. Olsen, "Study: Students Unfazed by Piracy," CNET news.com, 16 Sep 03.

28. J. P. Gray, *Why Our Drug Laws Have Failed and What We Can Do About It: A Judicial Indictment of the War on Drugs* (Temple Univ. Press, Philadelphia, 2001).

29. Singapore recently hanged Australian drug trafficker Nguyen Truong Van. See C. Reson, "Singapore Executes Australian," CNN.com, 2 Dec 05.

30. M. Haas, *The Singapore Puzzle* (Praeger, Oxford, 1999).

31. A. Caplan, "Is Biomedical Research Dangerous to Pursue?" *Science* **303**, 1142 (2004).

32. J. Pontin, "The Loss of Biological Innocence: Advances in Biotech Present Dark Possibilities and an Editor's Dilemma," *Technology Review*, March/April 2006.

33. H. I. Miller and G. Conko, *The Frankenfood Myth: How Protest and Politics Threaten the Biotech Revolution* (Praeger, Oxford, 2004).

34. M. Williams, "The Knowledge," *Technology Review*, March/April 2006.

35. D. Kennedy, "Stem Cells, Redux," *Science* **303**, 1581 (2004).

36. C. T. Scott, "Chimeras in the Crosshairs," *Nat. Biotechnol.* **24**, 487 (2006).

37. S. Lansing, "Dangerous Campaign Against Somatic-Cell Nuclear Transfer," *San Francisco Chronicle*, 25 Aug 05.

38. N. Hettinger, "Patenting Life: Biotechnology, Intellectual Property, and Environmental Ethics," *Environmental Affairs* **22**, 277 (1995).

39. J. Whitfield, "Black Death's DNA," *Nature Science Update*, 4 Oct 01.

40. L. A. Cole, *The Anthrax Letters: A Medical Detective Story* (National Academy Press, Washington, D.C., 2003).

41. C. Lacy, "Smallpox: A Scare and a Test," *New York Times*, 19 Sep 04; L. M. Wein, "Got Toxic Milk?" *New York Times*, 30 May 05; L. K. Donohue, "Censoring Science Won't Make Us Safer," *Washington Post*, 26 Jun 05.

42. D. Kennedy, "Two Cultures," *Science* **299**, 1148 (2003).

43. E. Lichtblau, "Response to Terror: Rising Fears That What We Do Know Can Hurt Us," *Los Angeles Times*, 18 Nov 01; P. Cohen, "Recipes for Bioterror: Censoring Science," *New Scientist*, 18 Jan 03.

44. R. Shattuck, *Forbidden Knowledge: From Prometheus to Pornography* (Harcourt Brace, New York, 1996).

45. R. Kurtzweil and B. Joy, "Recipe for Destruction," *New York Times*, 17 Oct 05. This op-ed piece criticizes publication in GenBank of the genome sequence for the 1918 HN influenza virus.

46. H. Marukami, *Underground: The Tokyo Gas Attack and the Japanese Psyche* (Vintage, New York, 2001).

47. A. M. Winkler, *Life Under a Cloud: American Anxiety About the Atom* (Oxford Univ. Press, New York, 1993).

48. T. Shanker and D. E. Sanger, "White House Wants to Bury Pact Banning Tests of Nuclear Arms," *New York Times*, 7 Jul 01; C. Lynch, "U.N. Backs Human Cloning Ban," *Washington Post*, 9 Mar 05.

49. A. C. Clarke, *The Hammer of God* (Spectra, New York, 1994).

CHAPTER THREE

1. A. Einstein, J. Mayer, and J. Holmes, *Bite-Size Einstein: Quotations on Just About Everything from the Greatest Mind of the Twentieth Century* (St. Martin's Press, New York, 1996).

2. I. Stewart, *Does God Play Dice? The New Mathematics of Chaos* (Blackwell, Oxford, 2002).

3. J. Gleick, *Chaos: The Making of a New Science* (Penguin, New York, 1988).

4. M. Luby, *Pseudorandomness and Cryptographic Applications* (Princeton Univ. Press, Princeton, 1996).

5. J.-N. Kapferer, *Rumors: Uses, Interpretations, and Images* (Transaction Publishers, Somerset, N.J., 1990).

6. The "Documentary Hypothesis" of the Pentateuch's origin is credible but controversial. See J. Blenkinsopp, *The Pentateuch* (Doubleday, New York, 1992).

7. J. E. Moulder, "Power-Frequency Fields and Cancer," *Crit. Rev. Biomed. Eng.* **26**, 1 (1998). Professor Moulder maintains an extensive bibliography on the subject of power lines and cancer at the Medical College of Wisconsin. His Web site is no longer active, but its content, updated through September 20, 2007, is available in PDF form at http://large.stanford.edu/rbl/publications/crime/reference.

8. A. Levin, "Airways in the USA Are the Safest Ever," *USA Today*, 30 Jun 06.

9. W. J. Broad, "NASA Puts Shuttle Mission's Risk at 1 in 100," *New York Times*, 26 Jul 05.

10. F. Wu, "Tornadoes and Trailer Parks: A Statistical Correlation," *Annals of Improbable Research* **1**, 4 (1995).

11. S. A. Sandford, "Apples and Oranges—A Comparison," *Annals of Improbable Research* **1**, 3 (1995).

12. B. Vonnegut, "Chicken Plucking as a Measure of Tornado Speed," *Weatherwise Magazine*, October 1975, p. 217.

13. T. Sah, *Fundamentals of Solid State Electronics* (World, Singapore, 1991).

14. B. Schneider, *Applied Cryptography: Protocols, Algorithms, and Source Code in C* (Wiley, New York, 1995).

15. E. Cole, *Hiding in Plain Sight: Steganography and the Art of Covert Communication* (Wiley, New York, 2003).

16. R. B. Laughlin, *A Different Universe: Remaking Physics from the Bottom Down* (Basic Books, New York, 2005).

17. J. D. Venhoeven, A. H. Pendray, and W. E. Dauksch, "The Key Role of Impurities in Ancient Damascus Steel Blades," *JOM* **59**, 58 (1998).

18. S. Garfield, *Mauve: How One Man Invented a Color That Changed the World* (Norton, New York, 2002); L. F. Haber, *The Chemical Industry, 1900–1930: International Growth and Technological Change* (Clarendon Press, Oxford,1971). See also R. J. Baptista, "Spies and Dyes," http://www.colorantshistory.org.

19. D. Owen, *Copies in Seconds* (Simon and Schuster, New York, 2004); P. M. Borsenberger, *Organic Photoreceptors for Xerography* (Marcel Dekker, New York, 1998).

20. S. Nakamura, G. Fasol, and S. J. Pearton, *The Blue Laser Diode: The Complete Story* (Springer, Heidelberg, 2000); J. E. Brewer, *Flash Memory Technologies* (Wiley, New York, 2006); J. D. Gralla, and P. Gralla, *The Complete Idiot's Guide to Understanding Cloning* (Alpha Books, New York, 2004).

CHAPTER FOUR

1. D. Coakley, H. Greenspun, and G. C. Gerard, *The Day the MGM Grand Hotel Burned* (Lyle Stuart, New York, 1982).

2. R. Fox and S. T. Harker, *Why You Lose at Poker* (ConJelCo Books, Pittsburgh, Pa., 2006).

3. J. von Neumann et al., *Theory of Games and Economic Behavior*, Commemorative Edition (Princeton Univ. Press, Princeton, 2004).

4. R. Miski, *Save Thousands Buying Your Next Car: Confessions of a Former Car Salesman* (RJM Publishing, Firsco, Tex., 2006).

5. M. Lessinger, *The Book of Bluffs: How to Bluff and Win at Poker* (Warner, New York, 2005).

6. L. R. Schreiber, *Poker as Life: 101 Lessons from the World's Greatest Game* (Hearst, New York, 2005).

7. This is a restatement of the widely (but not universally) accepted "subjective" theory of value. See C. Menger, J. Dingwall, and B. F. Hoselitz, *Principles of Economics* (New York Univ. Press, New York, 1981).

8. K. I. Furman, "Acute Vitaminosis A in an Adult," *Am. J. of Clin. Nutrition* **26**, 575 (1973).

9. C. Dreifus, "The Cyber-Maxims of Esther Dyson," *New York Times Magazine*, 7 Jul 06. See also E. Dyson, *Release 2.1* (Bantam, New York, 1998); M. Kanellos, "Gates: Restricting IP Rights Is Tantamount to Communism," CNET News.com, 6 Jan 05; J. E. Rogan, "The Morality of Intellectual Property Rights," *Religion and Liberty* **13**, 1 (Acton Institute, January/February 2003). See also P. Matier and A. Rose, "Governor's Republican Forgiveness Finds Some Room on the Bench," *San Francisco Chronicle*, 2 Aug 06.

10. *Economic Espionage Act of 1996,* United States Code, Title 18, Sections 1831–1839.

11. D. M. Hunt, *O. J. Simpson Facts and Fictions: News Rituals in the Construction of Reality* (Cambridge Univ. Press, Cambridge, 1999).

12. M. Aurelius and M. Staniforth, *Meditations* (Penguin, New York, 2002).

13. M. Friedman, "The Business Community's Suicidal Impulse," *Cato Policy Report* **21**, 2 (March/April 1999).

14. K. Austin, *World War 3.0: Microsoft and Its Enemies* (Random House, New York, 2001). See also S. A. Orlowski, "Microsoft Monopoly Says Apple Monopoly Is Too Restrictive," *The Register*, 20 Oct 03; and L. Lerer, "Symantec Snaps at Microsoft," *Forbes,* October 2006; A. Greenspan, "Antitrust," in A. Rand, *Capitalism: The Unknown Ideal* (New American Library, New York, 1967).

15. A. I. Poltorak and P. J. Lerner, *Essentials of Intellectual Property* (Wiley, New York, 2002).

16. A. A. Lipscomb and A. E. Bergh, *The Writings of Thomas Jefferson* (Thomas Jefferson Memorial Association, Washington, D.C., 1905), 13: 333, 17: 448.

17. J. J. Ellis, *American Sphinx: The Character of Thomas Jefferson* (Vintage, New York, 1997).

18. W. H. Pierson, *Jefferson at Monticello: The Private Life of Thomas Jefferson from Entirely New Materials* (Books for Libraries Press, Freeport, N.Y., 1862), pp. 103–111.

19. See http://www.launchpoker.com/poker-series/presidential _poker.

CHAPTER FIVE

1. L. Lessing, *Man of High Fidelity: Edwin Howard Armstrong* (Bantam, New York, 1969); E. I. Schwartz, *The Last Lone Inventor: A Tale of Genius, Deceit, and the Birth of Television* (Harper Collins, New York, 2002); H. F. Judson, "No Nobel Prize for Whining," *New York Times*, 20 Oct 03.

2. K. S. Kaplan et al., "System and Method for Creating Multiple Files from a Single Source File," U.S. Patent 6,594,674, 15 Jul 03.

3. *Diamond v. Diehr*, 450 U.S. 175 (1981), Docket No. 79–1112. This ruling contains the phrase: "Excluded from such patent protection are laws of nature, physical phenomena and abstract ideas."

4. S. Levgren, "One-fifth of Human Genes Have Been Patented, Study Reveals," *National Geographic News*, 13 Oct 05; R. M. Cook-Deegan and S. J. McCormack, "Patents, Secrecy and DNA," *Science* **293**, 217 (2001); E. Marshall, "DuPont Ups the Ante on Harvard's OncoMouse," *Science* **296**, 1212 (2002).

5. *Gottschalk v. Benson*, 409 U.S. 63 (1972), Docket No. 71–485. This ruling contains the phrase: "An algorithm, or

mathematical formula, is like a law of nature, which cannot be subject to a patent."

6. *Diamond v. Diehr.*

7. D. Diamond, "Gravity-Activated Fluid Displacement Power Generator," U.S. Patent 3,934,964, 27 Jan 76; E. T. Hartman, "Permanent Magnet Propulsion System," U.S. Patent 4,215,330; D. Baker, "Magnetic Propulsion Device," U.S. Patent 4,074,153, 14 Feb 79; H. R. Johnson, "Permanent Magnet Motor," U.S. Patent 4,151,431, 24 Apr 79; C. J. Flynn, "Methods for Controlling the Path of Magnetic Flux from a Permanent Magnet and Devices Incorporating the Same," U.S. Patent 6,246,561, 12 Jun 01; S. L. Patrick et al., "Motionless Electromagnetic Generator," U.S. Patent 6,362,718, 26 Mar 02; W. A. Green, "Piston Driven Rotary Engine," U.S. Patent 6,526,925, 4 Mar 03.

8. B. Volfson, "Space Vehicle Propelled by the Pressure of Inflationary Vacuum State," U.S. Patent 6,960,975, 1 Nov 05.

9. D. C. James, "Method for Data Compression," U.S. Patent 5,533,051, 2 Jul 06; C. E. Shannon and W. Weaver, *The Mathematical Theory of Communication* (Univ. of Illinois Press, Chicago, 1999).

10. R. B. Hartman, "Motorized Ice Cream Cone," U.S. Patent 5,971,829, 26 Oct 99; M. Jean-Prats, "Pillow with Retractable Umbrella," U.S. Patent 6,711,769, 30 Mar 04; A. Clyburg, "Inclining Coffin," U.S. Patent 6,725,510, 27 Apr 04; T. K. Amiss and M. H. Abbott, "Method for Exercising a Cat," U.S. Patent 5,443,036, 27 Aug 95; M. S. Brock, "Doggie Poop Freeze Wand," U.S. Patent 6,883,462, 26 Apr 05; G. B. Blonsky and C. E. Blonsky, "Apparatus for Facilitating the Birth of a Child by Centrifugal Force," U.S. Patent 3,216,423, 9 Nov 65; S. Olsen, "Method for Swinging on a Swing," U.S. Patent 6,368,227, 9 Apr 02.

11. J. Randi, *An Encyclopedia of Claims, Frauds, and Hoaxes of the Occult and Supernatural* (St. Martin's Griffin, New York, 1997).

12. C. Krol, "Cuckoos and Cocoa Puffs," *Skeptical Eye* **8**, 3 (1995); P. Cohen, "Poof! You're a Skeptic: The Amazing Randi's Vanishing Humbug," *New York Times*, 17 Feb 01.

13. C. Petit et al., "Mutated Polynucleotide Corresponding to a Mutation Responsible for Prelingual Non-syndromic Deafness in the Connexion 26 Gene and Method of Detecting This Hereditary Defect," U.S. Patent 5,998,147, 7 Dec 99; C. Petit et al., "Mutation Within the Connexion 26 Gene Responsible for Prelingual Non-syndromic Deafness and Method of Detection," U.S. Patent 5,998,147, 26 Nov 02; M. Crowley, "That's Outrageous: They Own Your Body," *Reader's Digest*, August 2006.

14. M. A. Katzer, "Model Train Control System," U.S. Patent 6,530,329, 11 Mar 03; C. Babcock, "Stakes Small, But Outcome Huge in Model Railroad Software Fight," *Information Week*, 6 Sep 06; B. Perents, "The Monster Arrives: Software Patent Lawsuits Against Open Source Developers," Technocrat.net, 30 Jun 06.

15. A. S. Green, "Homeland Security Agents Visit Toy Store," *Oregonian*, 28 Oct 04.

16. A. Orlowski, "Microsoft Blocks Cloners from Streaming Media Format," *Register*, 7 Jun 00; S. D. Levi et al., "Active Stream Format for Holding Multiple Media Streams," U.S. Patent 6,041,345, 21 Mar 00.

17. P. Davidson, "Patents Out of Control?" *USA Today*, 13 Jan 04.

18. J. Palmer, "Good Idea, But Where Are the Pizzas and Beer?" http://www.josephpalmer.com, 8 May 06.

19. M. Broersma, "Eolas Wins Round vs. MS in Browser Patent Fight," *eWeek*, 29 Sep 05; M. D. Doyle et al., "Distributed Hypermedia Methods for Automatically Invoking External Application Providing Interaction and Display of Embedded Objects with a Hypermedia Document," U.S. Patent 5,838,906, 17 Nov 98. After further litigation through appeal, all of which

upheld the judgment against Microsoft, Eolas and Microsoft settled in August 2007 for an undisclosed amount. See T. Bishop, "High-Profile, 8-Year Patent Dispute Settled," *Seattle Post Intelligencer*, 30 Aug 07; Bloomberg News, "Microsoft Settles a Dispute over a Feature in Its Browser," *New York Times*, 31 Aug 07.

20. T. Smith, "RIM Settles NTP Lawsuit for $450 M," *Register*, 16 Mar 05; I. Austin and L. Guernsey, "A Payday for Patents 'R' Us," *New York Times*, 2 May 05; T. J. Campana Jr. et al., "Electronic Mail System with RF Communication to Mobile Processors," U.S. Patent 6,317,592, 13 Nov 01; T. J. Campana Jr. et al., "Electronic Mail System with RF Communication to Mobile Processors," U.S. Patent 6,067,451, 23 May 00; T. J. Campana Jr. et al., "Electronic Mail System with RF Communications to Mobile Radios," U.S. Patent 5,819,172, 6 Oct 98; T. J. Campana Jr. et al., "Electronic Mail System with RF Communications to Mobile Processors Originating from Outside the Electronic Mail System and Methods of Operation Thereof," U.S. Patent 5,438,611, 1 Aug 95; T. J. Campana Jr., "System for Wireless Transmission and Receiving of Information and Method of Operation Thereof," U.S. Patent 6,272,190, 7 Aug 01; T. J. Campana Jr., "System for Wireless Serial Transmission of Encoded Information," U.S. Patent 5,751,773, 12 May 95; T. J. Campana Jr., "System for Wireless Transmission and Receiving of Information and Method of Operation Thereof," U.S. Patent 5,745,532, 28 Apr 98; T. J. Campana Jr., "System for Transferring Information from a RF Receiver to a Processor Under Control of a Program Stored by the Processor and Operation Thereof," U.S. Patent 5,631,946, 20 May 97; T. J. Campana Jr., "Electronic Mail System with RF Communications to Mobile Processors," U.S. Patent 5,625,670, 29 Apr 97.

21. A. S. Mutschler and E. Sperling, "Rambus Awarded $306.5 Million in Hynix Patent Trial," ElectronicsWeekly.com, 25 Apr 06; G. Gross, "Hynix Gets Delay in Rambus Patent Lawsuit," *InfoWorld*, 23 Aug 06; J. Rosenblatt, "Rambus Wins Hynix Patent

Case, Shares Rise 39 Percent (Update 7)," Bloomberg.com, 26 Mar 08; "Likely Appeal Deflates Rambus Patent Verdict News, Stock Drops," Entrepreneur.com, 28 Mar 08.

22. D. Cullen, "Toshiba Puts Lexar Royalties Spat to Bed," *Channel Register*, 15 Sep 06; P. Estakhri, "Flash Memory Leveling Architecture Having No External Latch," U.S. Patent 6,040,997, 21 Mar 00; P. Estakhri, "Data Pipelining Method and Apparatus for Memory Control Circuit," U.S. Patent 6,334,337, 16 Apr 02; P. Estakhri and B. Iman, "Increasing the Memory Performance of Flash Memory Devices by Writing Sectors Simultaneously to Multiple Flash Memory Devices," U.S. Patent 6,081,878, 27 Jun 00; P. Estakhri and B. Iman, "Organization of Blocks with a Non-volatile Memory Unit to Effectively Decrease Sector Write Operation Time," U.S. Patent 6,141,249, 31 Oct 00; P. Estakhri et al., "Moving Sectors Within a Block of Information in a Flash Memory Mass Storage Architecture," U.S. Patent 5,907,856, 25 May 99; P. Estakhri et al., "Space Management for Managing High Capacity Nonvolatile Memory," U.S. Patent 6,034,897, 7 Mar 00.

23. D. P. Hamilton, "Genentech Faces $500 Million Charge After Large Punitive-Damage Award," *Wall Street Journal*, 24 Jun 02; K. Itakura and D. Riggs, "Recombinant Cloning Vehicle Microbial Polypeptide Expression," U.S. Patent 4,704,362, 3 Nov 87.

24. S. Lohr, "I.B.M. Sues Amazon.com Over Patents," *New York Times*, 24 Oct 06. This suit later settled out of court for an undisclosed amount. See E. Morphy, "Amazon, IBM Patent Settlement Leaves Thorny Questions Unanswered," *Ecommerce Times*, 8 May 07.

25. J. Lettice, "Use Linux and You Will Be Sued, Ballmer Tells Governments," *Register*, 18 Nov 04.

26. W. J. Stilling, "Patent Term Extensions and Restoration Under the Hatch-Waxman Act," *Findlaw*, 20 Nov 02.

27. J. M. Mohr, "Legal Forum: Drug Product, Active Ingredient and Patent Term," ipFrontline, 6 May 06.

28. J. Gleick, "Patently Absurd," *New York Times Magazine*, 12 Mar 00.

29. H. J. Shaw and M. J. F. Digonnet, "Fiber Optic Amplifier," U.S. Patent 4,859,016, 22 Aug 89; D. Payne et al., "Erbium-Doped Fibre Amplifier with Shaped Spectral Gain," U.S. Patent 5,260,823, 9 Nov 93; T. Ergodan et al., "Laser Pumping of Erbium Amplifier," U.S. Patent 5,563,732, 8 Oct 96.

30. R. M. Schumacher and J. E. Matthews, "Structured Document Browser," U.S. Patent 5,933,841, 3 Aug 99; R. M. Schumacher and J. E. Matthews, U.S. Patent 6,442,574, 27 Aug 02; G. Gross, "Patently Outrageous? SBC Claims Patent on Framelike Browsing," *Computerworld*, 22 Jan 03.

31. Acacia Technologies' streaming media patents were upheld in 2003 but have received setbacks in the courts since then. See J. Leetzing, "Famed Patent Firm Acacia Hits Unwelcome Milestone," MarketWatch, 16 Nov 07; G. Daily, "Court Upholds Claims Construction Findings Against Acacia Technologies," streamingmedia.com, 17 Jul 06; E. Stasik, "IPTV Patent War: All Your Streams Belong to Us," *IP Television*, September 2005, p. 59; T. Riordan, "A Patent Owner Claims to Be Owed Royalties on Much of Internet's Media Content," *New York Times*, 16 Aug 04; J. Boreland, "Broad Patents on Streaming Media Upheld," CNET News.com,16 Jul 03; P. Yurt and H. L. Brown, "Audio and Video Transmission and Receiving System," U.S. Patent 5,132,992, 21 Jul 02; P. Yurt and H. L. Brown, U.S. Patent 5,253,275, 12 Oct 93; P. Yurt and H. L. Brown, U.S. Patent 5,550,863, 27 Aug 96; P. Yurt and H. L. Brown, U.S. Patent 6,002,720, 14 Dec 99; P. Yurt and H. L. Brown, U.S. Patent 6,144,702, 7 Nov 00. See also A. Sharma, "Acacia Subsidiary Acquires Rights to Patent for Video-conferencing Technology," TMCNet.com, 28 Mar 08; J. N. Hoover, "Acacia Research, Linux Patent Adversary, Has Long Litigation History," *Information Week*, 12 Oct 07; N. Anderson, "Acacia Claims Patent on CD Hyperlinks, Sues for Billions," Ars Technica, 18 Apr 07.

32. T. C. Wendt, "System and Method for Providing Integrated Voice, Video and Data to Customers Over a Single Network," U.S. Patent 7,075,919, 11 Jul 06.

33. R. Stross, "Why Bill Gates Wants 3,000 New Patents," *New York Times*, 31 Jul 05; E. J. Voetberg et al., "Method and System for Selection and Conjugating a Verb," U.S. Patent Application 20060195313, 31 Aug 06; R. K. Kott et al., "Systems and Methods to Facilitate Self-Regulation of Social Networks Through Trading and Gift Exchange," U.S. Patent Application 20060190281, 24 Aug 06.

34. A. Orlowski, "Cingular Applies to Patent Smileys," *Register*, 27 Jan 06; I. Cunningham and C. White, "Using Emotions, Such as for Wireless Devices," U.S. Patent Application 20060015812, 19 Jan 06.

35. *Diamond v. Chakrabarty*, 447 U.S. 303 (1980). This decision quotes the phrase "everything under the sun made by man," from S. Rep. No. 1979, 82d Cong., 2d Sess., 5 (1952). See also B. Kahin, "How Washington Reinvented Invention," CNET News.com, 5 Apr 06.

36. P. Hartman et al., "Method and System for Placing a Purchase Order Via a Communications Network," U.S. Patent 5,960,411; S. Shulman, "Software Patents Tangle the Web," *Technology Review*, 1 Dec 02.

37. R. L. Alcorn et al., "Internet Education Support System and Methods," U.S. Patent 6,988,138, 17 Jan 06.

38. T. Somers, "Review of Stem Cell Patents Is Sought," *San Diego Union-Tribune*, 19 Jul 06; J. A. Thomson, "Primate Embryonic Stem Cells," U.S. Patent 5,843,780, 1 Dec 98; J. A. Thomson, "Primate Embryonic Stem Cells," U.S. Patent 6,200,806, 13 Mar 01.

39. S. Albainy-Jones, "Submarine Patent That Surfaced After 24 Years Cost Monsanto $100 Million," *Patent Baristas*, 28 Feb 06; W. L. Miller et al., "Bovine Growth Hormone," U.S. Patent 6,692,941, 17 Feb 04; E. Marshall, "Biotech Giants Butt Heads

Over Cancer Drug," *Science* **288**, 2303 (2000); D. B. Ring, "Antigen-Binding Sites of Antibody Molecules Specific for Cancer Antigens," U.S. Patent 6,054,561, 25 Apr 00.

40. S. L. Stirland, "US Patent Reform: Could 2007 Be the Year?" *Intellectual Property Watch*, 25 Sep 06.

41. J. Leyden, "Case Against Dmitry Sklyarov Dropped," *Register*, 14 Dec 01; A. Newitz, "DVD Jon Lands Dream Job Stateside," *Wired News*, 18 Oct 05.

CHAPTER SIX

1. S. Turner, *Caging the Nuclear Genie: An American Challenge for Global Security* (Westview Press, Boulder, 1997).

2. V. E. Wagner, "Wie ist der Stand beim Ausstieg aus der Kernenergie in Schweden?" *Energie-Fakten*, 30 Jun 05.

3. L. I. Dorman, *Cosmic Rays in the Earth's Atmosphere and Underground* (Springer, Heidelberg, 2004).

4. I. Asimov, *The Explosions Within Us* (Ace, New York, 1976).

5. Letter from Attorney General John Ashcroft to the Honorable J. Dennis Hastert, Speaker of the House, 15 Oct 02. See "Report to Congress on Unauthorized Disclosure of Classified Information," at http://www.fas.org/sgp.

6. A. DeVolpi et al., *Born Secret: The H-Bomb, the Progressive Case, and National Security* (Pergamon Press, New York, 1981); John Aristotle Phillips, *Mushroom: The Story of the A-Bomb Kid* (Pocket Books, New York, 1978); J. Sterngold, "Nuclear Scientist Set Free After Plea in Secrets Case," *New York Times*, 14 Sep 00.

7. E. Freedman and A.-M. Murphy, "After 25 Years: U.S. v. The Progressive Inc. and Prior Restraint in the Era of the War on Terrorism," Proceedings of the 2004 Annual Conference of the Association for Education in Journalism and Mass Communication, Toronto, CA, Week 1. See http://list.msu.edu/cgi-bin/wa?A2=ind0411a&L=aejmc&T=0&P=10466.

8. H. Morland, "The H-Bomb Secret: How We Got It, Why We're Telling It," *The Progressive*, November 1979, p. 3; H. Morland, *The Secret That Exploded* (Random House, New York, 1981). See also *United States v. The Progressive*, 467 F. Supp. 990 (1979). The injunction against publication of the Morland article was issued by Judge Robert W. Warren of the U.S. District Court of the Eastern District of Wisconsin on March 26, 1979.

9. D. Stober, "C. Hansen, Collected Nuclear Arms Data," *San Jose Mercury News*, 1 Apr 03.

10. C. Hansen, *U.S. Nuclear Weapons: The Secret History* (Orion, New York, 1988).

11. R. W. Warren, *United States of America v. Progressive, Inc., Erwin Kroll, Samuel Day, Jr., and Howard Morland,* United States District Court, Western District of Wisconsin, 467 F. Supp 990, 28 Mar 79.

12. P. Collins, "The A-Bomb Kid," *Village Voice*, 17–23 Dec 03.

13. K. Davidson, "U.S. Questions UC's Running of Nuclear Lab," *San Francisco Chronicle*, 3 Jan 03; K. Chang, "Los Alamos Missing Secret Data," *New York Times*, 10 Jul 04; K. Davidson, "LANL Suspends 19 During Security Probe," *San Francisco Chronicle*, 23 Jul 04; S. Blakeslee, "Nuclear Lab's Missing Disks May Not Exist," *New York Times*, 12 Aug 04; K. Chang, "Government Penalizes University Overseer of Los Alamos Lab," *New York Times*, 29 Jan 05; R. Vartabedian, "Los Alamos Confirms Data Breach," *Los Angeles Times*, 26 Oct 06.

14. B. L. Holian, "Is There Really a Cowboy Culture at Los Alamos?" *Physics Today*, December 2004, p. 60.

15. W. Safire, "The Deadliest Download," *New York Times*, 29 Apr 99.

16. A. Specter, "Reports on the Cases of Dr. Wen Ho Lee and Dr. Peter Lee," *U.S. Congressional Record*, 20 Dec 01, pp. S13772–S13830. Senator Specter read into the *Record* a summary of a longer document titled "Report on Oversight of the

Wen Ho Lee Case," which was not approved by the Subcommittee on Department of Justice Oversight and was thus not an official statement of the Judiciary Committee. See http://www.fas.org/irp/congress/2001_rpt.

17. M. Purdy and J. Sterngold, "The Prosecution Unravels: The Case of Wen Ho Lee," *New York Times*, 5 Feb 01; S. S. Schwartz, "Scientist, Fisherman, Gardener . . . Spy?" *Bulletin of the Atomic Scientists* **56**, 31 (2000).

18. W. H. Lee, "Factual Basis for Plea of Dr. Wen Ho Lee," http://www.fas.org/irp/ops/ci.

19. V. Loeb, "No Bail in Atomic Data Case," *Washington Post*, 30 Dec 99.

20. J. Dahlkampf, G. Mascolo, and H. Stark, "Network of Death on Trial," Spiegel Online, 13 Mar 06; O. Mayer-Rüth, "Iranischer Physiker belastet mutmaßlichen deutschen Atomspion schwer," *Bayrischer Rundfunk*, 8 May 06; S. Coll, "Atomic Emporium," *New Yorker*, 7 Aug 07.

21. W. Fey, "Atomschmuggel: Reintaler vor Gericht," NZZ Online, 28 Dec 06. While this book was in press, a mistrial was declared and a new trial ordered. No trial date had been set as of 18 Apr 08. See "New Trial Ordered for German Accused of Aiding Libyan Nuclear Program," *Intl. Herald Tribune*, 14 Dec 07.

22. F. Abrams, *Speaking Freely: Trials of the First Amendment* (Viking, New York, 2005).

23. G. R. Stone, *Perilous Times: Free Speech in Wartime from the Sedition Act of 1798 to the War on Terror* (W. W. Norton, New York, 2004); F. W. Winterbotham, *The Ultra Secret* (Harper and Row, New York, 1974); P. J. Westwick, "In the Beginning: The Origin of Nuclear Secrecy," *Bulletin of the Atomic Scientists* **56**, 43 (2000).

24. J. B. Tucker, *Toxic Threat: Assessing Terrorist Use of Chemical and Biological Weapons* (MIT Press, Cambridge, 2000); J. B. Tucker, *War of Nerves: Chemical Warfare from World War I to al-Qaeda* (Pantheon, New York, 2006); M. Ignatieff, *The Lesser*

Evil: Political Ethics in an Age of Terror (Princeton Univ. Press, Princeton, 2004).

25. G. Allison, *Nuclear Terrorism: The Ultimate Preventable Catastrophe* (Henry Holt, New York, 2004).

Chapter Seven

1. D. R. Prothero, *Bringing Fossils to Life: An Introduction to Paleobiology* (McGraw-Hill, New York, 2003).

2. E. O. Wilson, *The Future of Life* (Knopf, New York, 2002).

3. R. Cook, *Vector* (Berkeley, New York, 1999).

4. M. Crichton, *Jurassic Park* (Knopf, New York, 1990); R. Preston, *The Cobra Event* (Random House, New York, 1998); P. B. Sammon, *Future Noir: The Making of Blade Runner* (Harper, New York, 1996).

5. H. G. Wells, *The Island of Dr. Moreau* (Modern Library, New York, 1996); M. M. Smith, *Spares* (Harper Collins, New York, 1996); N. Farmer, *The House of Scorpions* (Atheneum Books, New York, 2003).

6. C. N. Stewart et al., "Genetic Transformation, Recovery and Characterization of a Fertile Soybean Transgenic for a Synthetic *Bacillus Thuringiensis* cryIAc Gene," *Plant Physiology* **112**, 121 (1996); F. J. Perlack et al., "Insect Resistant Cotton Plants," *Nature Biotechnology* **8**, 939 (1990); M. Koziel et al., "Field Performance of Elite Transgenic Maize Plants Expressing an Insecticidal Protein Derived from *Bacillus Thuringiensis*," *Nature Biotechnology* **11**, 194 (1993); J. J. Estruch et al., "Transgenic Plants: An Emerging Approach to Pest Control," *Nature Biotechnology* **15**, 137 (1997).

7. A. Pellgrineschi et al., "Stress-Induced Expression in Wheat of the Arabidopsis Thaliana DREB1A Gene Delays Water Stress Symptoms Under Greenhouse Conditions," *Genome* **47**, 493 (2004); S. J. Oh et al., "Arabdopsis CBF3/DEB1A and ABF3 in Transgenic Rice Increase Tolerance to Abiotic Stress Without

Stunting Growth," *Plant Physiology* **138**, 341 (2005); M. Kasuga et al., "A Combination of the Arabidopsis DRGB1A Gene and Stress-Inducable rfd29A Promoter Improved Drought and Low-temperature Stress Tolerance in Tobacco by Gene Transfer," *Plant Cell Physiology* **45**, 346 (2004).

8. H. X. Zhang et al., "Engineering Salt-Tolerant Brassica Plant: Characterization of Yield and Seed Oil Quality in Transgenic Plants with Increased Vacuolar Sodium Assimilation," *Proc. Natl. Acad. Sci.* **98**, 12832 (2001).

9. J. Marshall, "Transgenic Pigs Are Rich in Healthy Fats," *New Scientist*, 27 Mar 06; L. Lai et al., "Generation of Cloned Transgenic Pigs Rich in Omega-3 Fatty Acids," *Nature Biotechnology* **24**, 435 (2006).

10. C. K. Yoon, "Altered Salmon Leading Way to Dinner Plate, But Rules Lag," *New York Times*, 1 May 00; M. Kaufman, "Frankenfish or Tomorrow's Dinner? Biotech Salmon Face a Current of Environmental Worry," *Washington Post*, 17 Oct 00.

11. J. C. Rapp et al., "Biologically Active Human Interferon? Produced in the Egg White of Transgenic Hens," *Transgenic Research* **12**, 569 (2003); W. Velandez, H. Lubon, and W. Drohan, "Transgenic Livestock as Drug Factories," *Scientific American*, January 1997, p. 55.

12. A. Pollack, "Kraft Recalls Taco Shells with Bioengineered Corn," *New York Times*, 23 Sep 00; A. Pollack, "Safeway Recalls Taco Shells After Test Questions Corn Origin," *New York Times*, 12 Oct 00; A. Pollack, "Altered Corn Surfaced Earlier," *New York Times*, 4 Sep 01; N. Boyce, "Taco Trouble," *New Scientist*, 7 Oct 00; J. Kluger, "Tempest in a Taco Shell," *Time*, 2 Oct 00; J. Kaiser, "Panel Urges Further Study of Biotech Corn," *Science* **290**, 1867 (2000).

13. A.-M. Chèvre et al., "Gene Flow from Transgenic Crops," *Nature* **389**, 924 (1997); P. W. Hendrick, "Invasion of Transgenes from Salmon or Other Genetically Modified Organisms into Natural Populations," *Canadian Journal of*

Fisheries and Aquatic Sciences **58**, 841 (2001); A. A. Snow, "Transgenic Crops—Why Gene Flow Matters," *Nature Biotechnology* **20**, 542 (2002); A. Daniell, "Molecular Strategies for Gene Containment in Transgenic Crops," *Nature Biotechnology* **20**, 581 (2002); A. Ritala et al., "Measuring Gene Flow in the Cultivation of Transgenic Barley," *Crop Science* **42**, 278 (2002); L. S. Watrud et al., "Evidence of Landscape-Level, Pollen-Mediated Gene Flow from Genetically Modified Creeping Bentgrass with CP4 EPSPS as a Marker," *Proc. Natl. Acad. Sci.* **101**, 14533 (2004).

14. J. E. Losey, L. S. Rayor, and M. E. Carter, "Transgenic Pollen Harms Monarch Larvae," *Nature* **399**, 214 (1999); Z. Y. Huang et al., "Field and Semifield Evaluation of Impact of Transgenic Canola Pollen on Survival and Development of Worker Honey Bees," *Journal of Economic Entomology* **97**, 1517 (2004); L. A. Morandin and M. L. Winston, "Effects of Novel Pesticides on Bumble Bee (*Hymmenoptera: Apidae*) Colony Health and Foraging Ability," *Environmental Entomology* **32**, 555 (2003).

15. J. Couzin, "U.S. Agencies Unveil Plan for Biosecurity Peer Review," *Science* **303**, 1595 (2004); D. A. Shea, "Balancing Scientific Publication and National Security Concerns: Issues for Congress," *CRS Report* RL31695; A. Froelich, "Washington Watch: To Publish, or Not to Publish, after 9/11," *BioScience*, January 2003; R. M. Gregg, "Rogue Science," *Georgetown Law Journal*, August 2003; M. Clayton, "Academia Becomes Target of New Security Laws," *Christian Science Monitor*, 24 Sep 02; D. MacKenzie and S. P. Westphal, "Should the Genetic Sequences of Deadly Diseases be Kept Secret?" *New Scientist*, 20 Jul 02; C. L. Epstein, "Controlling Biological Warfare Threats," *Critical Reviews in Microbiology* **27**, 321 (2001); J. Couzin, "A Call for Restraint on Biological Data," *Science* **297**, 749 (2002).

16. C. Hogg, "Taiwan Breeds Green-Glowing Pigs," BBC News, 12 Jan 06.

17. An excellent introduction to microbiological warfare issues can be found on the Web site of Washington State University Emeritus Professor Ronald E. Hurlbert. See http://www.slic2.wsu.edu:82/hurlbert/micr0101/index.html.

18. P. J. Boyer, "The Ames Strain," *New Yorker*, 12 Nov 01; J. Miller, W. Broad and S. Engelberg, *Germs* (Simon and Schuster, New York, 2002); K. Alibeck and S. Handelman, *Biohazard* (Random House, New York, 1999).

19. J. B. Tucker, *Scourge: The Once and Future Threat of Smallpox* (Grove Press, New York, 2001).

20. T. O'Toole, M. Mair, and T. V. Inglesby, "Shining Light on 'Dark Winter,'" *Clinical Infectious Diseases* **34**, 972 (2002); M. Fish, "Officials: Bioterror Would Challenge Health Care Facilities," CNN.com, 16 Jan 02.

21. A. McCook, "PNAS Publishes Bioterror Paper, After All," *Scientist* **6** (29 Jun 05).

22. J. Parkhill et al., "Genome Sequence of *Yersinia pestis,* The Causative Agent of Plague," *Nature* **413**, 523 (2001); T. D. Read et al., "The Genome Sequence of *Bacillus anthracis* Ames and Comparison to Closely Related Bacteria," *Nature* **423**, 81 (2003); J. J. Esposito et al. "Genome Sequence Diversity and Clues to the Evolution of Variola (Smallpox) Virus," *Science* **313**, 807 (2006); A. M. Rosengard et al., "Variola Virus Immune Evasion Design: Expression of a Highly Efficient Inhibitor of Human Complement," *Proc. Natl. Acad. Sci.* **99**, 8808 (2002); R. Seghadri et al., "Complete Sequence of Q-fever Pathogen *Coxiella burnetti,*" *Proc. Natl. Acad. Sci.* **100**, 5455 (2003).

23. J. S. Driscoll, *Antiviral Drugs* (Wiley, New York, 2005); D. D. Richman, *Antiviral Drug Resistance* (Wiley, New York, 1996).

24. R. J. Jackson et al., "Expression of Mouse Interleukin-4 by a Recombinant Ectromelia Virus Suppresses Lymphocyte Responses and Overcomes Genetic Resistance to Mousepox," *J. Virol.* **75**, 1205 (2001).

25. W. Whitehouse, "DNA Databases 'No Use to Terrorists,'" BBC News, 15 Jan 03.

26. J. Cello, A. V. Paul, and E. Wimmer, "Chemical Synthesis of Poliovirus cDNA: Generation of Infectious Virus in the Absence of Natural Template," *Science* **27**, 1016 (2002); S. Bloch, "A Not-So-Cheap Stunt," *Science* **297**, 769 (2002); R. Weiss, "Polio-causing Virus Created in N.Y. Lab: Made-From-Scratch Pathogen Prompts Concerns About Bioethics, Bioterrorism," *Washington Post*, 12 Jul 02. See also F. Gottron, "Synthetic Poliovirus: Bioterrorism and Science Policy Implications," *CRS Report* RS21369.

27. J. Rifkin, *The Biotech Century* (Penguin Putnam, New York, 1999); O. Morton, "Biology's New Forbidden Fruit," *New York Times*, 11 Feb 05.

28. S. G. Stolberg, "Transition in Washington: Research and Morality; Stem Cell Advocates in Limbo," *New York Times*, 20 Jan 01; N. Wade, "Scientists Divided on Limit of Federal Stem Cell Money," *New York Times*, 16 Aug 01; N. Wade, "Grants for Stem Cells Are Delayed," *New York Times*, 24 Apr 01; W. Safire, "Stem Cell Genie," *New York Times*, 10 Jun 01; S. G. Stolberg, "House Judiciary Committee Panel Passes No-Clone Bill," *New York Times*, 25 Jul 01; F. Bruni, "Bush Says He Will Veto Any Bill Broadening His Stem Cell Policy," *New York Times*, 14 Aug 01; G. Niebuhr, "Religions Ponder Stem Cell Issues," *New York Times*, 24 Aug 01; S. G. Stolberg, "Bush Denounces Cloning and Calls for Ban," *New York Times*, 27 Nov 01; S. G. Stolberg, "Controversy Reignites Over Stem Cells and Clones," *New York Times*, 18 Dec 01; S. G. Stolberg, "Dispute Over Cloning Experiments Intensifies," *New York Times*, 6 Mar 02; M. J. Sandel, "The Anti-Cloning Conundrum," *New York Times*, 28 May 02; N. Wade, "Word War Breaks Out In Research on Stem Cells," *New York Times*, 21 Dec 02; L. R. Kass, "How One Clone Leads to Another," *New York Times*, 24 Jan 03; D. D. Kirkpatrick, "Bush Defends Stem-Cell Limit, Despite Pressure Since Reagan

Death," *New York Times*, 16 Jun 04; S. G. Stolberg, "In Rare Threat, Bush Vows Veto of Stem Cell Bill," *New York Times*, 21 May 05.

29. M. Herper, "Japan's Stem-Cell Bid Lures U.S. Researchers," *Forbes*, 4 Mar 02; D. Kocieniewski, "McGreevy Signs Bill Creating Stem Cell Research Center," *New York Times*, 13 May 04; E. Rosenthal, "Britain Embraces Embryonic Stem Cell Research," *New York Times*, 24 Aug 04; W. Safire, "California's Stem Cell Gold Rush," *New York Times*, 15 Dec 04; M. McIntire, "Fearing New York Might Fall Behind, Senator Proposes Stem Cell Institute," *New York Times*, 17 Jan 05; P. Bulluck, "Massachusetts Legislators Endorse Study of Stem Cells," 1 Apr 05; G. Reuthling, "Illinois to Pay for Stem Cell Research," *New York Times*, 13 Jul 05.

30. R. McGuirk, "Australia Lifts Ban on Cloning Embryos," *Boston Globe*, 7 Dec 06; R. Cohen, "House Votes to Lift Stem Cell Ban but Bush Veto Certain," *New Jersey Star-Ledger*, 12 Jan 07; D. D. Bilefsky, "EU to Fund Stem Cells," *International Herald Tribune*, 24 Jul 06; P. N. Spotts, "Stem-Cell Research Surges Ahead of Lawmakers," *Christian Science Monitor*, 20 May 05; A. Thomas, "World-Leading Stem Cell Science Centre Built in Edinburgh," *Christian Today*, 13 Jan 07.

31. S. Levin, "Pitt Researcher Warned Work Will Be Scrutinized," *Pittsburgh Post-Gazette*, 12 Jan 07; C. Dreifus, "At Harvard's Stem Cell Center, the Barriers Run Deep and Wide," *New York Times*, 24 Jan 06; M. Gazzaniga, "All Clones Are Not the Same," *New York Times*, 16 Feb 06.

32. A. Pollack, "Rebuilding with Stem Cells," *New York Times*, 30 May 00; N. Wade, "Teaching the Body to Heal Itself; Work on Cells' Signals Fosters Talk of a New Medicine," *New York Times*, 7 Nov 00; A. Pollack, "Scientists Seek Ways to Rebuild the Body, Bypassing the Embryos," *New York Times*, 18 Dec 01; N. Wade and S. G. Stolberg, "Scientists Herald Versatile Adult Cell," *New York Times*, 25 Jan 02.

33. K. H. S. Campbell et al., "Sheep Cloned by Nuclear Transfer from a Cultured Cell Line," *Nature* **380**, 64 (1996); I. Wilmot et al., "Viable Offspring Derived from Fetal and Adult Mammalian Cells," *Nature* **385**, 810 (1997).

34. B. C. Lee et al., "Dogs Cloned from Adult Somatic Cells," *Nature* **436**, 641 (2005); Q. Zhou et al., "Generation of Fertile Cloned Rats by Regulating Oocytic Activation," *Science* **302**, 179 (2003); R. P. Lanza et al., "Extension of Cell Life-Span and Telmoere Length in Animals Cloned from Senescent Somatic Cells," *Science* **288**, 605 (2000); T. Shen et al., "A Cat Cloned by Nuclear Transplantation," *Nature* **415**, 859 (2002); C. Galli et al., "A Cloned Horse Born to Its Dam Twin," *Nature* **415**, 859 (2002); I. A. Polejaeva et al., "Cloned Pigs Produced by Nuclear Transfer from Adult Somatic Cells," *Nature* **407**, 86 (2000); G. Kolata, "What Is Warm and Fuzzy Forever? With Cloning, Kitty," *New York Times*, 15 Feb 02. See also http://www.savingsandclone.com.

35. I. Oransky, "All Hwang Human Cloning Work Fraudulent," *Scientist*, 10 Jan 06; D. Kennedy, "Editorial Retraction," *Science* **311**, 335 (2006). The fraudulent claims in the notorious Hwang Woo-seok case had mainly to do with stem cell generation. The original paper claiming successful somatic nuclear transfer was not retracted. See W. S. Hwang et al., "Evidence of the Pluripotent Human Embryonic Stem Cell Line Derived from a Cloned Bastocyst," *Science* **303**, 1669 (2004).

36. A. Pollack, "Missing Limb? Salamander May Have Answer," *New York Times*, 24 Sep 02.

37. G. Struhl, "A Homeotic Mutation Transforming Leg to Antenna in Drosophila," *Nature* **292**, 635 (1981); L. W. Browder, C. A. Erickson, and W. R. Jeffery, *Developmental Biology*, 3rd ed. (Saunders, Philadelphia, 1991).

38. L. Moss, "Scientists Say Disease Research at Risk After 'Knee-Jerk' Reaction to Embryo Proposals," *Scotsman*, 7 Jan 07; J. Wheldon, "Scientists Denounce Plans to Outlaw 'Chimera' Embryo Experiments," *Daily Mail*, 4 Jan 07.

39. See, for example, J. Schienda et al., "Somitic Origin of Limb Muscle Satellite and Side Population Cells," *Proc. Natl. Acad. Sci.* **103**, 945 (2006).

40. T. L. Roth et al., "Survival of Sheep x Goat Hybrid Inner Cell Masses After Injection into Ovine Embryos," *Biology of Reproduction* **41**, 675 (1989); E. D. Zanjani et al., "Engraftment and Long-Term Expression of Human Fetal Hemopoietic Stem Cells in Sheep Following Transplantation in Utero," *Journal of Clinical Investigation* **89**, 1178 (1992); H. Nagashima et al., "Turtle-Chicken Chimera: An Experimental Approach to Understanding Evolutionary Innovation in the Turtle," *Developmental Dynamics* **232**, 149 (2005); E. Balaban, M. A. Teillet, and N. LeDouarin, "Application of the Quali-Chock Chimera System to the Study of Brain Development and Behavior," *Science* **241**, 1339 (1988). See also R. Weiss, "Of Mice, Men and In-Between," *Washington Post*, 20 Nov 04.

41. R. Morelle, "Of Mice and Men," BBC News, 7 Nov 06; J. S. Orr, "Merging Animal, Human Cells—Where's the Line?" *San Francisco Chronicle*, 12 Mar 06; G. M. Lamb, "A Mix of Mice and Men," *Christian Science Monitor*, 23 Mar 05; C. Y. Johnson, "Blending of Species Raises Ethical Issues," *Boston Globe*, 19 Apr 05.

42. L. Marks, *Sexual Chemistry: A History of the Contraceptive Pill* (Yale Univ. Press, New Haven, 2001); C. Djerassi, *One Man's Pill: Reflections on the 50th Anniversary of the Pill* (Oxford Univ. Press, New York, 2004).

43. M. Duff, "Evolution Challenged in U.S. Schools," BBC News, 11 Mar 02; D. C. Dennett, *Darwin's Dangerous Idea* (Simon and Schuster, New York, 1996); W. Irvine, *Apes, Angels, and Victorians: The Story of Darwin, Huxley, and Evolution* (McGraw-Hill, New York, 1995); A. M. Gunn, *Intelligent Design and Fundamentalist Opposition to Evolution* (McFarland, Jefferson, N.C., 2006); J. G. West, "Idea Not Based on Religion," *USA Today*, 21 Dec 05; A. L. Melott, "Intelligent Design Is Creationism in a Cheap Tuxedo," *Physics Today* **55**, 48 (2002).

CHAPTER EIGHT

1. See http://www.biofact.com/cloning/jokes.html.

2. E. S. Lander et al., "Initial Sequencing and Analysis of the Human Genome," *Nature* **409**, 860 (2001); E. S. Lander et al., "International Human Genome Sequencing Consortium: Finishing the Euchromatic Sequence of the Human Genome," *Nature* **431**, 931 (2004).

CHAPTER NINE

1. M. McLuhan, E. McLuhan, and R. Zingrone, *Essential McLuhan* (Basic Books, New York, 1995).

2. W. M. Baum, *Understanding Behaviorism: Behavior, Culture and Evolution* (Blackwell, Oxford, 2005); M. R. Rosenzweig, S. M. Breedlove, and N. V. Watson, *Biological Psychology: An Introduction to Behavioral and Cognitive Neuroscience* (Sinauer Associates, Sunderland, Mass., 2005); J. A. Mills, *Control: A History of Behavioral Psychology* (New York Univ. Press, New York, 2000); B. F. Skinner, *Beyond Freedom and Dignity* (Hackett, Indianapolis, 2002).

3. H. G. Wells was a fiery scientific idealist and proponent of a utopia of "perpetual criticism, increase and more knowledge and more." See the transcript of his address to the Australian public, transmitted through ABC on 19 Jan 39 (Wells Collection, Univ. of Illinois). See also H. G. Wells, *Men Like Gods* (Kessinger, Whitefish, Mont., 2005); F. Bacon, "New Atlantis," in *Famous New Deals of History,* ed. by H. E. Barnes and C. M. Andrews (Kessinger, 2007); J. Spargo, "When H. G. Wells Smokes the Opium of Utopia," *New York Times,* 12 Dec 20; R. Barbrook, *Imaginary Futures: From Thinking Machines to the Global Village* (Pluto Press, 2007).

4. See the Spam Store at http://www.spam.com.

5. L. A. Canter and M. S. Siegel, *How to Make a Fortune on the Information Superhighway: Everyone's Guide to Marketing on the Internet and Other On-Line Services* (HarperCollins, New York, 1995).

6. K. Hafner, *Where Wizards Stay Up Late: The Origins of the Internet* (Simon and Schuster, New York, 1998).

7. R. Grover et al., "Can Mad Ave. Make Zap-Proof Ads?" *Business Week*, 2 Feb 04; F. Moore, "Stealth Advertising to Foil Ad Zappers," CBS News, 10 Jan 03; T. Lefton, "You Can't Stop These Ads," *Computerworld*, 21 Mar 01; S. Woolley, "Zap!" *Forbes*, 29 Sep 03; D. Kiley, "Television: Counting the Eyeballs," *Business Week*, 16 Jul 06; G. Goodale, "Life After the 30-Second Advertising Spot," *Christian Science Monitor*, 21 Jul 06; D. Lieberman and L. Petrucca, "New TiVo Service to Measure Its Ad-Zapping Fallout," *USA Today*, 26 Jul 06.

CHAPTER TEN

1. This does not include laws that require legislated education benefits to be distributed without discrimination. See *Plyer v. Doe*, 457 U.S. 202 (1982); and C. R. Sunstein, *The Second Bill of Rights: FDR's Revolution and Why We Need It More Than Ever* (Perseus, New York, 2006).

INDEX